A FEEDBACK CONTROL MECHANISM FOR THE AUTOMATION OF STRESS AND LOAD TESTING

by

Mohamad S. Bayan

APPROVED BY SUPERVISORY COMMITTEE:

____

João W. Cangussu, Chair

____

Gopal Gupta

____

Kendra M.L. Cooper

____

Kamil Sarac

I would like to dedicate this Doctoral dissertation to my family, especially to Mom and Dad for instilling the importance of hard work and higher education and to Najwa for her patience and support.

A FEEDBACK CONTROL MECHANISM FOR THE AUTOMATION OF STRESS AND LOAD TESTING

by

MOHAMAD S. BAYAN, B.S., M.S.

DISSERTATION

Presented to the Faculty of

The University of Texas at Dallas

in Partial Fulfillment

of the Requirements

for the Degree of

DOCTOR OF PHILOSOPHY IN COMPUTER SCIENCE

THE UNIVERSITY OF TEXAS AT DALLAS

August, 2010

UMI Number: 3421491

ProQuest LLC
789 East Eisenhower Parkway
P.O. Box 1346
Ann Arbor, MI 48106-1346

ACKNOWLEDGEMENTS

First and foremost, special thanks go to my advisor, Dr. João W. Cangussu, for his support and guidance throughout the program. Dr. Cangussu helped me identify this topic and worked with me to overcome challenges and avoid possible dead ends. It has been very rewarding to have him as my advisor and I admire his willingness to put students' interests first. Without his support this dissertation would not have been in its final form.

My gratitude also goes to Dr. Kendra Cooper who has given me many suggestions, comments, and help on my research. I would also like to thank my committee members for serving on my dissertation committee.

I owe further gratitude to my parents, my wife, and other family members for their patience, support, and encouragement during the past several years as without their support this achievement would not have been possible.

April, 2010

A FEEDBACK CONTROL MECHANISM FOR THE AUTOMATION OF STRESS AND LOAD TESTING

Publication No. ____________________

Mohamad S. Bayan, Ph.D.
The University of Texas at Dallas, 2010

Supervising Professor: João W. Cangussu

Stress and Load Testing are of special interest to many segments of the software industry. However, existing approaches do not provide a solution that is automated, widely applicable, flexible, efficient, accurate, and based on a rigorous theoretical foundation. For example, while some approaches are automated, they cannot be used to stress/load test different types of applications. A novel Stress and Load testing approach is proposed here. It is the first approach to use a proportional integral derivative (PID) Controller to automatically drive the inputs and to achieve a pre-specified level of stress/load for resources of interest. A new adaptive System Identification algorithm is proposed that automatically identifies inputs that impact the resources. A new automatic PID Tuning algorithm is also presented that defines the internal parameters (K_P, K_I, and K_D) that are important to achieve a stable process. The PID Tuning can be done with different goals in mind depending on the Rise Time, Overshoot, Settling Time, and Error at Equilibrium. The approach is rigorously validated with four experiments in addition to multiple simulation runs. The results indicate the new automated approach offers a solution that significantly advances the state-of-the-research in Stress and Load Testing.

TABLE OF CONTENTS

ACKNOWLEDGEMENTS . . . v

ABSTRACT . . . vi

LIST OF TABLES . . . x

LIST OF FIGURES . . . xi

CHAPTER 1. INTRODUCTION . . . 1

1.1 Motivation . . . 1

1.2 Goals . . . 2

1.3 Overview of The Approach . . . 4

1.4 Contributions . . . 6

1.5 Structure . . . 7

CHAPTER 2. INPUT IDENTIFICATION . . . 8

2.1 System Identification . . . 8

2.2 SI Based Algorithm for Input Identification . . . 13

2.3 Adaptive SI Based Algorithm for Input Identification . . . 14

2.4 Validation of Automatic Input Identification . . . 16

2.5 Cut Off Value Analysis . . . 22

2.6 Related Work: Feature Selection . . . 22

2.7 Input Identification vs. Feature Selection . . . 27

2.8 Conclusion . . . 29

CHAPTER 3. CONTROLLER . . . 30

3.1 PID Controller . . . 30

3.2 Using PID Controller in the proposed approach . . . 32

3.3 Advantages of Using a Controller . . . 34

3.4 Comparison with the Binary Approach . . . 35

3.5 Conclusion . . . 36

CHAPTER 4. AUTOMATIC CONTROLLER TUNING 38
4.1 Ziegler-Nichols Tuning Method 40
4.1.1 Ziegler-Nichols: Closed Loop Tuning Method 42
4.1.2 Ziegler-Nichols: Open Loop Tuning Method (Reaction Curve) 42
4.2 Automatic Controller Tuning for Software Applications 45
4.3 Conclusion 46

CHAPTER 5. EXPERIMENTAL RESULTS 50
5.1 Experiment α: Memory Usage 52
5.2 Experiment β: Response Time 54
5.2.1 Controlling one variable 56
5.2.2 Controlling one variable of two inputs 58
5.2.3 Controlling two variables of two inputs 60
5.3 Experiment γ: Zip Experiment 61
5.3.1 Identifying Sensitive Inputs 63
5.3.2 Tuning the PID Controller 63
5.3.3 Controlling the Zip Application 64
5.4 Experiment δ: Web Application 66
5.4.1 Tuning the PID Controller 69
5.4.2 Controlling One Type of Client 69
5.4.3 Controlling two Types of Clients 73
5.4.4 Controlling Three Types of Clients 74
5.4.5 Controlling Four Types of Clients 77
5.4.6 Controlling Seven Types of Clients 78
5.5 Conclusion 80

CHAPTER 6. RELATED WORK 82
6.1 Toward A Structural Load Testing Tool 82
6.2 Detecting Buffer Overflow 83
6.3 Deterministic Markov State Testing 85
6.4 Stress Testing Real-Time Systems with Genetic Algorithms 86
6.5 Stress Testing of Multimedia Systems 86
6.6 Stress/Load Testing tools 88
6.7 Conclusion 90

CHAPTER 7. CONCLUSION AND FUTURE WORK 92
7.1 Conclusion 92
7.2 Future Work 93

REFERENCES 96

VITA

LIST OF TABLES

2.1 Number of required test cases for the system identification and the brute-force approach with $c = 3$. 21

2.2 Feature selection Example 25

4.1 Effects of increasing a control Parameter (K_P, K_I, or K_D) independently on the Control Loop behavioral parameters 41

4.2 Ziegler-Nichols Tuning Method: Closed Loop 42

4.3 Ziegler-Nichols Tuning Method: Open Loop Reaction Curve 44

5.1 Inputs to the Zip Application. 62

5.2 Results of Applying the System identification Procedure to the Zip application random test cases. 64

5.3 Web Application Setup - Controlling 1 type of clients. 71

5.4 Web Application Setup - Controlling 2 types of clients. 73

5.5 Web Application Setup - Controlling 3 types of clients. 75

5.6 Web Application Setup - Controlling 4 types of clients. 77

5.7 Web Application Setup - Controlling 7 types of clients. 79

6.1 Comparison of Stress/Load Testing tools by Functionalities 89

6.2 Comparing the proposed work with other approaches 91

LIST OF FIGURES

1.1 Proposed Approach 4

2.1 Pseudo-code for the adaptive resource sensitive input identification procedure 15

2.2 Distribution of inputs affecting the resource of interest for the simulation runs with a total of 15 inputs. 17

2.3 Polynomial Curve Fitting 18

2.4 Results of the Automatic Identification of inputs for test sets with size $|T| = 250, 300, \ldots, 1000, 1050$. [18] 19

2.5 Results of the improved approach of the Automatic Identification of inputs for test sets with size $|T| = 250, 300, \ldots, 1000, 1050$. 20

2.6 Comparison between the results of the propose approach using the system identification technique against the brute-force algorithm with $c = 3$. 20

2.7 Results for different cut off values. 23

2.8 A taxonomy of feature selection algorithms 26

2.9 Summary of feature selection methods 27

2.10 Feature Selection vs. Input Identification 28

3.1 PID Controller 30

3.2 Cruise Control System. 32

3.3 Instance of a controller used in Figure 1.1. 33

3.4 Application of Binary Approach to generate test cases to achieve a pre-defined level of response time for a client server system. 36

4.1 Typical Pid Tuning Graph 39

4.2 Pid Controller Common Performance Criteria - Overshoot, Rise Time, and Settling Time 41

4.3 Closed Loop Tuning - After increasing the proportional gain, the process loop rises and then dies down 43

4.4 Closed Loop Tuning - After increasing the proportional gain, the process loop oscillates 43

4.5 Open Loop Tuning - Reaction Curve 44

4.6 Pseudo-code for the Automatic Tuning algorithm 47

4.7 Pseudo-code for the function representing the exponential model. 48

4.8 Pseudo-code for the simulateData function 48

5.1 Results of applying a PID Controller to achieve a specified level of memory usage without oscillations. 51

5.2 Results of applying a PID Controller to achieve a specified level of memory usage with oscillations. 51

5.3 Results of applying a PID Controller to potentially identify memory leaks. . . . 53

5.4 Application of a PID Controller to achieve a certain average response time for the remote client server application. Only regular clients are considered and the total number of clients is the only input variable used by the controller. . 56

5.5 Stability Graph. 57

5.6 Application of a PID Controller to achieve a certain average response time for the remote client server application. Both Regular and VIP clients are considered and the total number of clients is the only input variable used by the controller. 59

5.7 Total number of clients and number of Regular and VIP clients used in Figure 5.6. The percentage of regular clients is randomly determined. 59

5.8 Application of a PID Controller to achieve a certain average response time for the remote client server application. Both Regular and VIP clients are considered; total number of clients and percentage of Regular Clients are both used by the controller. 60

5.9 Total number of clients and number of Regular and VIP clients used in Figure 5.8. The percentage of regular clients is determined by the controller. 61

5.10 Tuning the controller for the Zip Experiment - Determining K_P 65

5.11 Tuning the controller for the Zip Experiment - Determining K_I and K_D 65

5.12 Results of experiment γ where the three input variables "Size", "numFiles", and "CompRatio" are used to control response time. 66

5.13 Web Application Hardware Setup 67

5.14 Web Application Software Setup 68

5.15 Web Service Overview 69

5.16 Tuning the controller for the Web Application Experiment - Determining K_P . 70

5.17 Tuning the controller for the Web Application Experiment - Determining K_I and K_D 70

5.18 Results of controlling one type of client (Type1) to achieve a 5000 ms response time. 72

5.19 Total number of Type1 clients used in Figure 5.18. 72

5.20 Results of controlling two types of clients (Type1 and Type50) to achieve a 5000 ms response time. 73

5.21 Total number of clients and number of Type1 and Type50 clients used in Figure 5.20. 74

5.22 Results of controlling three types of clients (Type1, Type50 and Type500) to achieve a 5000 ms response time. 75

5.23 Total number of clients and number of Type1, Type50, and Type500 clients used in Figure 5.22. 76

5.24 Results of controlling four types of clients (Type1, Type50, Type100 and Type500) to achieve a 5000 ms response time. 77

5.25 Total number of clients and number of Type1, Type50, Type100, and Type500 clients used in Figure 5.24. 78

5.26 Results of controlling all seven types of clients to achieve a 5000 ms response time. 79

5.27 Total number of clients and number of all seven types of clients used in Figure 5.26. 80

6.1 Example of a load sensitive fault in an object-oriented program 84

6.2 Grosso's test case generation process 85

CHAPTER 1

INTRODUCTION

1.1 Motivation

Software testing is composed of numerous specialized testing activities, including functional testing (e.g., unit, integration, system) and performance testing (e.g., load, stress). Load and stress testing are important for many segments of the software industry including embedded systems, web applications, and database applications. For example, embedded systems have a constrained environment with limited system resources, making it important to assess an embedded system's operation when loaded or stressed. Also, it is important to assess the number of requests that web applications and database applications can handle within a specific response time. Some systems may execute successfully when it is executed for a short time or under a small load but may fail when executed under a heavy load or over a long period of time. Recently, the Federal Government issued a "Cash For Appliances" rebate program where each state is given a set amount of funds for energy-efficient appliances. Residents of the state of Texas had the options to call a toll-free phone lines or to go to a web site to reserve a rebate starting on Wednesday April 7th 2010 at 7 am. However, tens of thousands of frustrated residents were unable to get through the toll-free phone lines or the web site because of busy signals and blank Web pages for hours [9]. The Web site was shut down for two and half hours before coming back online to limited traffic. This is a very good example of the importance of stress and load testing.

Load Testing [20, 21] assesses how a system performs under a given "load", which is the rate at which transactions are submitted to the system [19]. One of Load Testing's objectives is to determine the maximum sustainable load the system can handle. Load testing reveals programming errors that would not appear if the system is executed with a small workload.

Such errors, called load-sensitive faults, emerge when the system is executed under a heavy load.

Stress Testing [20, 21, 19] requires subjecting a system to an unreasonable load with the intention of breaking the system. A Stress Test denies a system the resources (e.g., RAM, disc, interrupts) needed to process a certain load. Stress tests are designed to cause a failure. Stress Testing tests the system's fault recovery capability. The system is not expected to process the overload without adequate resources but to behave (e.g., fail) in a decent manner (e.g., not corrupting or losing data). In other words, the desired result is a graceful degradation leading to a non-catastrophic failure.

Existing Stress and load testing approaches present limitations with respect to automation and applicability. Even though some approaches present an automated solution, they lack in terms of applicability where they are applicable to a specific type of software applications such as multimedia systems [49] and telecommunication systems [15, 16]. Some approaches are restricted to the stress of only one resource like detecting buffer overflow [31, 32], limiting their scope and applicability. Existing approaches don't offer the capability to maintain a specific load level for a period of time regardless of the dynamic changes in the testing environment and this is important since some load-sensitive faults emerge only when the system is executed under a heavy load over a long period of time.

1.2 Goals

Given the limitations of currently available Load and Stress Testing approaches, there is a need for additional research in this area. The goal of this work is to present a Load and Stress Testing technique with the following characteristics:

- Automated: Automatically achieving a desired load level using only random test cases. The random test cases can be the same test cases used for functional testing.

- Widely Applicable: Applying the technique to a wide range of applications where its not limited by number or type of system inputs.
- Flexible: Applying the technique to a wide range of resources like memory usage, response time, or cpu usage.
- Efficient: Controlling how fast or slow the system under test achieves the desired load.
- Accurate: Achieving the desired load and maintaining that load for a period of time regardless of the testing environment.
- Based on Rigorous Theoretical Foundation: The technique is based on efficient rigorous theories like PID Control and System Identification.

In the above characteristics list, the first item is automation. Efficient rigorous theories are used to achieve automation. First, using System Identification, inputs sensitive to a specific resource like response time are identified automatically. Second, exploiting Control Theory, the system under test is controlled to automatically achieve a pre-specified level of stress/load for resources of interest. To achieve efficiency and accuracy, controller tuning theory is used to automatically tune the controller based on the stress/load testing goals.

The approach proposed here automates the complete process, applies to any system, and controls any desired resource. By achieving these goals and automatically driving the resource usage to its limit, load-sensitive faults can be detected and performance issues can be verified under stress conditions. Automatic identification of these faults can significantly improve the quality of released products as well as reduce required test effort. It improves the quality of the product where the system does not crash in conditions of insufficient resources such as memory or disk space, unusually high concurrency, or denial of service attacks. It also reduces required test effort and cost since the goal is to automate the stress/load testing process.

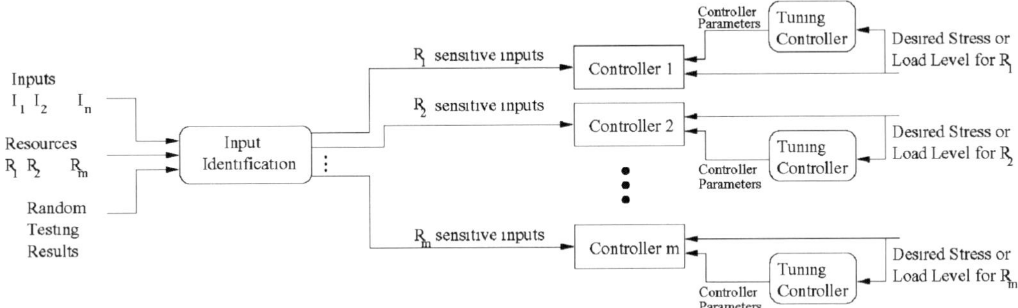

Figure 1.1. Proposed Approach

1.3 Overview of The Approach

Figure 1.1 presents the approach's general structure. The approach has three major components: "Input Identification," "Controller Tuning," and "Controller." The first component identifies the inputs that affect resources of interest. As described in more detail in Chapter 2, the effectiveness of Stress/Load Testing decreases considerably if not all the inputs that affect a resource are used to control the resource. Also, the use of all inputs will lead to an excessive number of test cases. Therefore, inputs that directly impact the resource need to be identified.

This identification is accomplished by the use of system-identification techniques [40]. System-identification techniques deal with the problem of building a model (mathematical relation) based on observations of a system's behavior when executing in a specific environment. To generate such a model, System-identification needs some random test data to start with which represents the system's behavior. System-identification does not rely on source code inspection which considerably increases the proposed approach's applicability. Automatically identifying inputs sensitive to a certain resource is the first step toward achieving the main goal of automating the stress/load testing process since the identified inputs are the one to be controlled by the Controller. System-identification is discussed in Chapter 2.

A major component of the proposed approach is the controller. Controllers are designed to eliminate the need for continuous operator attention. Cruise Control in a car and a house thermostat are common examples of how controllers are used to automatically adjust some variables to hold the measurement at a certain value like a certain speed or temperature. The proposed approach is based on the application of a feedback PID (Proportional, Integral, and Derivative) Controller to drive the input. This approach makes the system achieve a specified level of resource usage. For example, if the user defines that the system should be tested with a memory use of 95%, starting from an initial input value, the PID Controller automatically changes the input(s) until the desired level of stress has been achieved. Stress can also be simultaneously achieved for more than one resource, for example, memory and bandwidth. Experiments have shown that resources used by the PID Controller itself are almost negligible and have minor or null impact on systems resources. The Controller is the driver in the process of automating stress/load testing. The Controller is discussed in Chapter 3.

After the set of inputs that affect the resource have been identified, the "Controller Tuning" component tunes the controller. The tuning (as described in Chapter 4) can be done with different goals. For example, if a load is to be achieved but not exceeded, the controller must be tuned to converge to the desired load without overshooting. An automatic tuning technique is described in Section 4.2 to compute the chosen controller's internal parameters. Automatic Controller Tuning achieves the efficiency and the accuracy needed for the proposed approach.

Once the controller has been tuned, it can be used to drive the inputs to achieve selected levels of stress/load for a specified resource. Given an initial test case, the "Controller" computes the sequence of test cases to drive the resource to the desired load/stress level; discussed in Chapter 3. This task is, in general, performed manually and the approach proposed here automates this process.

Four experiments were conducted to show the approach's applicability. The first experiment shows that a PID Controller can be indeed used for both Stress Testing and Load Testing of one resource with one input variable. Additionally, analyzing changes in input and memory allocation allowed for detection of memory leaks that were injected into the source code. The second experiment shows that when more than one input significantly affects the resource, all the inputs must be controlled to achieve the desired level of stress/load. An efficient method is presented to identify these inputs, based on the sound theory of System Identification [40], in Chapter 2. The third experiment uses the Unix utility Zip to apply the approach to a widely used application. The fourth experiment is the most realistic example so far where the application under test is a Web Application hosted by a JBoss Server.

Controlling three different resources in the four experiments proves the flexibility of the approach. Automatic Controller Tuning algorithm was used in two of the experiments and proved the efficiency and accuracy of the approach by rapidly reaching the desired load level and maintaining it. Finally, controlling four different types of applications shows that the approach is widely applicable and not restricted.

1.4 Contributions

The work presented here provides a technique to automate the stress/load testing process. It is a widely applicable, flexible, accurate, and efficient technique which is based on rigorous theories. As part of achieving the goals of the proposed technique, the following four contributions are attained.

First, an Input Identification algorithm is proposed to identify inputs that are sensitive to a specific resource. The algorithm is based on System Identification techniques. For example, a system may have $\{I_1, I_2, \ldots, I_{15}\}$ as the set of inputs while only inputs $\{I_4, I_7, I_8, I_{11}\}$ significantly affect resource R_1. The Input Identification algorithm, using random test data, determines the subset of inputs that affect a certain resource. With simulation runs, the algorithm achieved an accuracy of 94%. Also, the application of the proposed algorithm to a software product is proved in an experiment.

The second contribution of the proposed work is applying the PID control theory to stress and load testing. To the best of my knowledge, the work proposed here is the first to use the PID controller to control and automate the process of stress/load testing a software application. PID Control theory is widely used in industrial control applications like regulation of temperature or speed. However, applying it to a software applications and especially to automate the process of stress or load testing a software application has not been done before.

The third main contribution of the proposed work is developing a PID tuning technique that is applicable to software applications. Tuning a PID controller is critical to its success in achieving the stress/load testing goals. Without PID tuning, the PID controller may not be able to control the software application to achieve the desired level of stress or load testing. To the best of my knowledge, the work proposed here is the first to develop a PID tuning technique applicable to software applications.

The fourth contribution is that the work proposed is applicable to all software applications as long as the application is properly defined in terms of inputs and outputs where random test data can be collected. The proposed work is also applicable to any resource as long as the resource is measurable. Also, as part of the contributions, four experiments were performed successfully to prove the applicability and accuracy of the proposed approach.

1.5 Structure

The proposed work is organized in the following order, a description of each major component, Chapter 2 presents the sensitive input identification component; Chapter 3 describes the control approach used to change the inputs and drive the resource to the desired level of stress/load; and Chapter 4 describes the importance of tuning a controller. Experimental results are presented in Chapter 5. Relevant existing methods for Stress and Load Testing are described in Chapter 6. This chapter also compares this approach to other existing approaches. A brief discussion of existing tools for Load Testing of web applications is presented in Chapter 6. Finally, Chapter 7 concludes the proposed work and discusses the future research plan.

CHAPTER 2
INPUT IDENTIFICATION

Before automatically driving a resource to the desired stress/load level, inputs that are sensitive to that specific resource must be identified. For example, a system may have $\{I_1, I_2, \ldots, I_{15}\}$ as the set of inputs while only inputs $\{I_4, I_7, I_8, I_{11}\}$ significantly affect resource R_1. As observed in Section 5.2, a resource cannot be controlled if an adequate set of inputs is not identified correctly. Assume that only inputs $\{I_4, I_7\}$ are used to perform stress testing. All other inputs assume random values. This approach's difficulty is that inputs I_8 and I_{11} may be needed to trigger some stress-related defects. The chance of finding such defects is very small if the inputs are not properly controlled. Alternatively, all the available inputs $\{I_1, I_2, \ldots, I_{15}\}$ could be controlled. However, this practice would lead to a larger test suite than when only the required inputs are used; controlling all available inputs will unnecessarily increase the controller's complexity.

2.1 System Identification

A model capturing a system's behavior must be available if such system is to be controlled. A model that can capture the dominant behavior of aspects of any system is unlikely to be statically created due to the variety of scenarios and environments where these systems execute. However, based on observations of a system's behavior when executing in a specific environment, a model can be dynamically customized for the relevant scenario. Ljung states "System Identification deals with the problem of building mathematical models of dynamical systems based on observed data from the system." [40] In other words, a mathematical relation between the measurements of the behavior of the system and the external influences (inputs to the system) is determined without worrying about what is happening inside the system. The ingredients to apply System Identification techniques are present and a model can be inferred from the observations.

According to Ljung, System Identification evolves around the following concepts [40]:

- Model m is a relationship between observed quantities where typically a mathematical expression is used to represent the relationship.
- True Description S is an abstraction of the object to be modeled as in most cases it is not realistic to achieve a "true" description.
- Model Class M is a collection of models. For example, M can be parameterized by a finite-dimensional parameter, like "all linear state-space models of order n".
- Complexity C is a measure of size or flexibility of a model class.
- Information obtained from the observed data and previous knowledge about the object to be modeled like a model class.
- Estimation is the process of choosing a model based on the information provided. Estimation Data, or training data, is the data used for selecting the model Z_e^N where N is the size of the data set.
- Validation, also known as generalization, is the process of ensuring that the chosen model is valid not only for the estimation data, but also for other data sets. The Data sets are called validation data Z_v.
- Model Fit $F(m, Z)$ is the measure of how well a particular model m fits to a particular data set Z.

$$\begin{cases} \dot{x}(t) = Ax(t) + Bu(t) \\ y(t) = Cx(t) + Du(t) \end{cases} \tag{2.1}$$

A number of techniques, such as Least-Square and Markov Parameters [40, 36] are available to identify state-space models. State space [26, 41] models are mathematical models of physical systems that use a set of input, output, and state variables to describe a system by a set of first-order differential equations. State space representation provides a convenient way to model systems with multiple inputs and outputs (MIMO). Eqn. 2.1 is a state space representation of a system with p inputs, q outputs, and n state variables with $A \in \mathbb{R}^{n \times n}$, $B \in \mathbb{R}^{n \times p}$, $C \in \mathbb{R}^{q \times n}$, and $D \in \mathbb{R}^{q \times p}$ and where $u(t) \in \mathbb{R}^p$ is the input of the system, $y(t) \in \mathbb{R}^q$ the output, $x(t) \in \mathbb{R}^n$ the state, and $\dot{x}(t) = \frac{d}{dt}x(t)$. The minimum number of state variables required to represent a given system is called the order of the model.

Let A, B, C, and D be the matrices in Eqn. 2.1 and organize them as indicated in Eqn. 2.2.

$$Y(t) = \begin{bmatrix} x(t+1) \\ y(t) \end{bmatrix} \quad \Theta = \begin{bmatrix} A & B \\ C & D \end{bmatrix} \quad \Phi(t) = \begin{bmatrix} x(t) \\ u(t) \end{bmatrix} \tag{2.2}$$

Using the definitions of $Y(t)$, Θ, and $\Phi(t)$ above, Eqn. 2.1 can be rewritten as Eqn. 2.3.

$$Y(t) = \Theta\Phi(t) \tag{2.3}$$

The observation of the system resources and the inputs/outputs of the application are used to construct the vectors $Y(t)$ and $\Phi(t)$. To verify the initial experiments' applicability and accuracy, the Least-Square identification procedure available in MATLAB was chosen; a concise description of this technique is presented below.

Least-Square [40] is a computational approach of fitting mathematical model to data. The best model (best-fit curve) is the model that has the minimal sum of the deviations squared (least square error) from the data. Least squares problems are divided into linear and non-linear problems. For example, suppose that the data set consists of m points $(x_1, y_1), (x_2, y_2), ..., (x_m, y_m)$ where x_i is the independent variable and y_i is the dependent variable. The fitting curve $f(x)$ has the residual (error) r from each data point $(r_1 = y_1 - f(x_1), r_2 = y_2 - f(x_2), ..., r_m = y_m - f(x_m))$. According to the method of least squares, it finds the best fitting curve when the sum, S in Eqn. 2.4, of squared residuals is a minimum.

$$S = r_1^{\,2} + r_2^{\,2} + ... + r_m^{\,2} = \sum_{i=1}^{m} r_i^{\,2} = \sum_{i=1}^{m} [y_i - f(x_i)]^2 \tag{2.4}$$

Consider a system of m linear equations in n unknown coefficients, $\beta_1, \beta_2, , \beta_n$, with $m > n$:

$$\sum_{j=1}^{n} X_{ij}\beta_j = y_i, \ (i = 1, 2, \ldots, m) \tag{2.5}$$

and in a matrix form as

$$\boldsymbol{X\beta} = \mathbf{y} \tag{2.6}$$

where

$$\boldsymbol{X} = \begin{pmatrix} X_{11} & X_{12} & \cdots & X_{1n} \\ X_{21} & X_{22} & \cdots & X_{2n} \\ \vdots & \vdots & \ddots & \vdots \\ X_{m1} & X_{m2} & \cdots & X_{mn} \end{pmatrix}, \quad \boldsymbol{\beta} = \begin{pmatrix} \beta_1 \\ \beta_2 \\ \vdots \\ \beta_n \end{pmatrix}, \quad \boldsymbol{y} = \begin{pmatrix} y_1 \\ y_2 \\ \vdots \\ y_m \end{pmatrix}. \tag{2.7}$$

Given a set of m data points $y_1, y_2, \ldots, y_m$, consisting of experimentally measured values taken at m values $x_1, x_2, \ldots, x_m$ of an independent variable, and given a model function $y = f(x, \boldsymbol{\beta})$, with $\boldsymbol{\beta} = (\beta_1, \beta_2, \ldots, \beta_n)$, it is desired to find the parameters β_j such that the model function fits "best" the data.

$$f(x, \boldsymbol{\beta}) = \sum_{j=1}^{n} \beta_j \phi_j(x). \tag{2.8}$$

The residual is the difference between the value of the dependent variable and the predicted value from the estimated model:

$$r_i(\boldsymbol{\beta}) = y_i - f(x_i, \boldsymbol{\beta}), \ (i = 1, 2, \ldots, m) \tag{2.9}$$

The sum of the squared residuals is:

$$S(\boldsymbol{\beta}) = \sum_{i=1}^{m} r_i^2(\boldsymbol{\beta}) \tag{2.10}$$

The summation S is minimized when its gradient with respect to each parameter is equal to zero. The elements of the gradient vector are the partial derivatives of S with respect to the parameters:

$$\frac{\partial S}{\partial \beta_j} = 2\sum_i r_i \frac{\partial r_i}{\partial \beta_j} = 0 \ (j = 1, 2, \ldots, n) \tag{2.11}$$

Since $r_i = y_i - \sum_{j=1}^{n} X_{ij}\beta_j$, the derivatives are

$$\frac{\partial r_i}{\partial \beta_j} = -X_{ij} \tag{2.12}$$

Substitution of the expressions for the residuals and the derivatives into the gradient equations gives

$$\frac{\partial S}{\partial \beta_j} = -2\sum_{i=1}^{m} X_{ij}\left(y_i - \sum_{k=1}^{n} X_{ik}\beta_k\right) = 0 \tag{2.13}$$

Upon rearrangement, the normal equations are obtained.

$$\sum_{i=1}^{m}\sum_{k=1}^{n} X_{ij}X_{ik}\hat{\beta}_k = \sum_{i=1}^{m} X_{ij}y_i \ (j = 1, 2, \ldots, n) \tag{2.14}$$

The normal equations are written in matrix notation as

$$(\boldsymbol{X}^\top \boldsymbol{X})\hat{\boldsymbol{\beta}} = \boldsymbol{X}^\top \boldsymbol{y} \tag{2.15}$$

The least square estimate of $\boldsymbol{\beta}$ is given by $\hat{\boldsymbol{\beta}}$:

$$\hat{\boldsymbol{\beta}} = (\boldsymbol{X}^\top \boldsymbol{X})^{-1}\boldsymbol{X}^\top \boldsymbol{y} \tag{2.16}$$

Finally, this Least-Square approach computes an approximation ($\hat{\Theta}_N^{LS}$) for the matrix (Θ) in Eqn. 2.3 as the following:

$$\hat{\Theta}_N^{LS} = \left[\frac{1}{N}\sum_{t=1}^{N} \Phi(t)\Phi^T(t)\right]^{-1} \frac{1}{N}\sum_{t=1}^{N} \Phi^T(t)Y(t) \tag{2.17}$$

2.2 SI Based Algorithm for Input Identification

The use of System Identification techniques [40] allows for identification of the subset of inputs that affect the output (the resource of interest). For explanation purposes, assume a system of order n with m inputs and only one output. In this case, the output part of Eqn. 2.1 is given by Eqn. 2.18

$$y(t) = [c_1\ c_2 \ldots c_n] \begin{bmatrix} x_1(t) \\ x_2(t) \\ \vdots \\ x_n(t) \end{bmatrix} + [d_1\ d_2 \ldots d_m] \begin{bmatrix} u_1(t) \\ u_2(t) \\ \vdots \\ u_m(t) \end{bmatrix} \tag{2.18}$$

The method described in Section 2.1 computes all the matrices A, B, C, and D of the system for a given input set and the associated output data. In this discussion, the interest is restricted to matrix D. The output variable $y(t)$ is the resource under consideration. The matrix D determines which inputs $\{u_1, u_2, \ldots, u_m\}$ affect the output. That is, if $d_i = 0$, then, the input has no direct impact on the resource to be controlled. However, using the value 0 when deciding whether to include the input leads to many false positives (The approach has selected inputs that do not actually affect output). To minimize this problem, a cut-off value is defined. First, matrix D is normalized to 100% with respect to the summation of its absolute values, as given in Eq. 2.19. Absolute values are used to accommodate inputs that affect the resource in either a positive or a negative manner.

$$\begin{aligned} d_j^n &= \frac{d_j \times 100}{\sum_{i=1}^{m} |d_i|} \quad \text{for } j = 1, 2, \ldots, m \\ D^n &= [d_1^n\ d_2^n \ldots d_m^n] \end{aligned} \tag{2.19}$$

Now, a cut-off value z is pre-specified to determine which inputs affect the output. The input u_i is considered to affect the output if $d_i^n > z$. Clearly, z's value impacts the approach's accuracy. A small value for z would lead to cases where the input is selected but should not be considered. A large value for z would lead to a dropping of inputs that should be considered. The impact the value of the cut-off parameter z has on the input identification procedure's accuracy is evaluated in Section 2.5. Results show that better accuracy is achieved when the value of z is approximately 1.5%.

2.3 Adaptive SI Based Algorithm for Input Identification

The approach proposed here is an improvement over the solution in the previous section (Results of both approaches presented in Section 2.4). Figure 2.1 presents the pseudo-code for the improved adaptive approach. In this approach, once the sensitivity of each input parameter is determined, this information is used to iteratively eliminate parameters that do not affect the output. That is, assume an initial number of test cases are randomly generated and used to compute the initial sensitivity of each input parameter. If n input parameters are considered, the system identification procedure uses a matrix of results of size $n+1$ (n inputs and 1 output) to produce the normalized sensitivity of each parameter $S_{I_1}, S_{I_2}, \ldots, S_{I_n}$. Now, any input with a sensitivity value less than the cut-off value is removed because it has negligible impact on the output (resource). Next, twenty additional test cases are randomly generated. The results are appended to the matrix. The system identification procedure is executed again with both the updated matrix and the remaining variables (the ones not eliminated in the previous iteration). This process is repeated until the total set of inputs impacting the outputs has not changed for four consecutive iterations. That is, 80 new test cases have been added but no input has been removed. This indicates that the remaining inputs highly impact the output because for the same set of inputs an increase in the number of test cases has not altered the selection of inputs.

As presented next in Section 2.4, Figure 2.1's Adaptive SI Based Algorithm provides better results than the SI Based Algorithm described in Section 2.2. This improvement is believed to be based on two factors. First, removing inputs that clearly do not impact a resource decreases the noise (unnecessary data) in the next iteration. This, consequently, improves accuracy. Second, improvement occurs because more useful information is added as additional test cases are appended to the test suite.

```
Inputs = {I1,I2,...,In}
SensitiveInputs = Inputs

// Generate 200 initial test cases
TestSuite = GenRandom(Inputs,200)

// Store the results of test cases in a matrix
Matrix_Results = Run(TestSuite)
count = 0

// Stoping criterion: 4 iterations whithout removing any input
// from the input set
while (count < 4)

   currentSensitiveInputsSize = Size(SensitiveInputs)
   // Applying the System identification Procedure
   Sensitivity = ApplySystemIdentification(Matrix_Results)
   for j=1 to Size(Inputs)
      if (Sensitivity(I(j))< cutoff_value)
         // Remove input Ij since it does not impact output
         SensitiveInputs = SensitiveInputs - I(j)
      end
   end

   // Generate 20 extra test cases changing only the sensitive inputs
   // all other inputs are kept constant
   NewTestCases = GenRandom(Inputs,20)

   //Append the new results to the matrix
   Matrix_Results = Append(Matrix_Results,Run(NewTestcases))

   // verifying if size of sensitiveInputs set has changed
   if (currentSensitiveInputsSize == Size(SensitiveInputs))
      count = count + 1
   else
      count = 1 ;
   end

end
```

Figure 2.1. Pseudo-code for the adaptive resource sensitive input identification procedure

2.4 Validation of Automatic Input Identification

In this section, simulation is used to show the general accuracy of the input identification procedure and to analyze the effects of the cut-off values. Also, the results of the SI Based Algorithm 2.2 is compared with the Adaptive SI Based Algorithm 2.3 and compare the approach with brute-force algorithm and Feature Selection algorithm; Application of the procedure to a software product is presented in Section 5.3.

The first issue to be considered for the input identification is the number of test cases, i.e., the size of a test set T needed to make a proper selection. Use of a brute-force algorithm demands an extremely large number of test cases when the number of inputs is large. Not only individual values but also all possible combinations of the inputs need to be evaluated to ensure the inputs affect the resource. Eqn. 2.20 provides the number of combinations required by a brute-force algorithm multiplied by a constant c. A single test case cannot show that an input or set of inputs affects the resource. At least three test cases ($c = 3$) are required: an initial test case t_1; a test case showing a decrease in the resource usage when compared to t_1; and a test case showing an increase in the resource usage when compared to t_1. The value of c, according to the experiments, will range from 5 to 10.

$$|T| = c \times \sum_{k=0}^{n} \binom{n}{k} = c \times 2^n \qquad (2.20)$$

Assume that 15 inputs are available for an application P. To verify the approach's accuracy, a series of simulation runs are executed for test sets in the range $|T| = 250, 300, \ldots, 1000, 1050$. For each size of test set, 100 simulation runs were executed. The number of inputs affecting the output is randomly decided. The distribution of inputs affecting output is shown in Figure 2.2. Up to 9 inputs can simultaneously affect output. But, on average, 4 inputs impact the output.

To generate simulation data, the next step is to define how the inputs will affect the output. Eqn. 2.21 defines how the resource y should behave with respect to changes in the

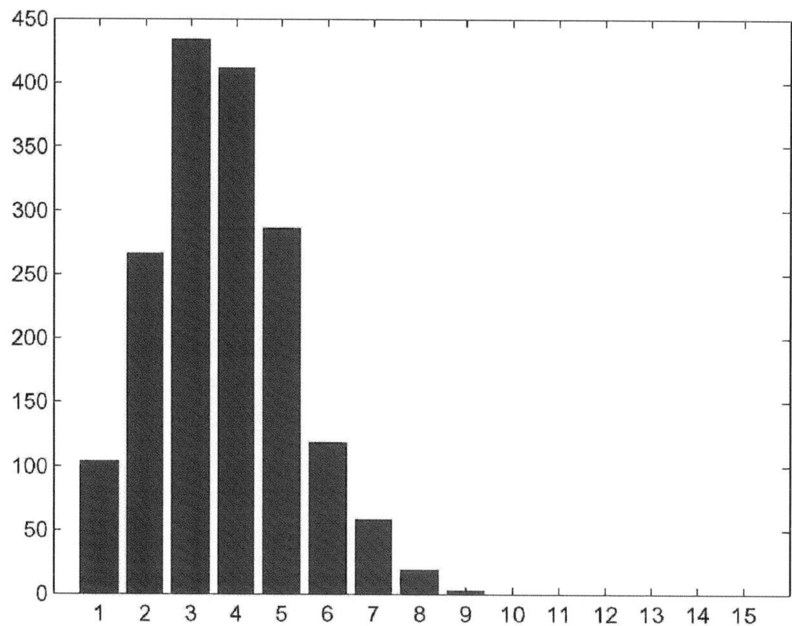

Figure 2.2. Distribution of inputs affecting the resource of interest for the simulation runs with a total of 15 inputs.

selected inputs.

$$y = \sum_{i=1}^{n} f_k(u_i) + w \tag{2.21}$$

where n is the number (randomly generated) of selected inputs ($n \leq 15$), w is random noise introduced in the function, and $f_k(u)$ is one of five available functions: $f_1(u) = \sqrt{u}$, $f_2(u) = a \times u$, $f_3(u) = u^2$, $f_4(u) = -b \times u$, or $f_5(u) = c_8u^8 + c_7u^7 + c_6u^6 + c_5u^5 + c_4u^4 + c_3u^3 + c_2u^2 + c_1u + c_0$. The values of constants a and b are generated randomly as well as each value of k. Also the range of values for each input u_i are defined randomly. The fifth function $f_5(u)$ is a polynomial function of degree 8 where the coefficients are generated randomly using Polynomial Curve Fitting. Polynomial Curve fitting is the process of constructing a polynomial curve that has the best fit to a series of data points. First, to generate the data points, two data vectors m and n are randomly generated where $minimum(m) = minimum(u_i)$, $maximum(m) = maximum(u_i)$, and $m_{j+1} - m_j = k$ where k is a randomly generated constant. Second, the coefficients of a polynomial that fits the data (m and n) in a least-squares sense are generated using Polynomial Curve Fitting available in MATLAB. For example, in Figure 2.3(a), m and n were randomly generated where $m = [31.0, 60.9, 90.8, 120.7, 150.6, 180.5, 210.4, 240.3, 270.2, 300.1, 330.0]$ and $n = [1513, 1578, 846, 1900, 1120, 1572, 212, 1805, 434, 217, 769]$. Then, using Polynomial Curve Fitting, the polynomial that fits the data is generated with coefficients $[0, 0, 0, 0, 0.4, -194.6, 4736.8]$. Another example is presented in Figure 2.3(b), where $m = [0, 86.3, 172.6, 258.9, 345.2, 431.5, 517.8, 604.1, 690.4, 776.7, 863.0]$ and $n = [1972, 284, 576, 1030, 952, 987, 792, 564, 298, 1347, 743]$ and the polynomial that fits the data is generated with coefficients $[0, 0, 0, 0, 0.2, -32.8, 1943.6]$. Eighth degree polynomial curve fitting is used to guarantee a complex mathematical relation between the inputs and the output.

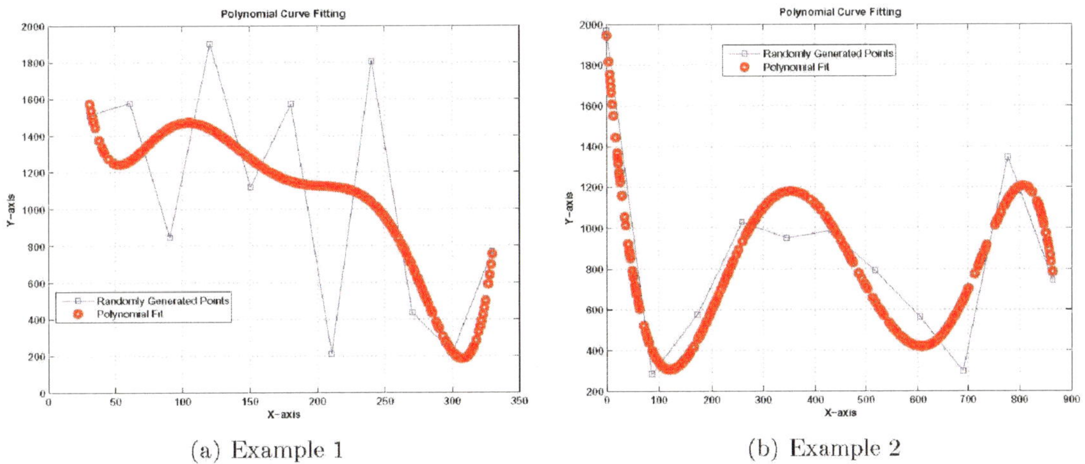

(a) Example 1 (b) Example 2

Figure 2.3. Polynomial Curve Fitting

A vector $U = [I_1\ I_2\ \ldots\ I_{15}]$ indicates the inputs used to compute the output y. $I_j = 0$ determines the input has no impact. $I_j = 1$ shows otherwise. After using the system identification procedure, a new vector $U^s = [I_1^s\ I_2^s\ \ldots\ I_{15}^s]$ determines, in the same way, which inputs the approach has selected. Now, let $U - U^s = [s_1\ s_2\ \ldots\ s_{15}]$. Success is represented by $s_j = 0$, a false positive by $s_j = -1$, and a false negative by $s_j = 1$.

Figure 2.4 depicts the results of the simulation runs for the SI Based Algorithm 2.2. The Figure shows a small increase in accuracy (an increase in the number of successes and a

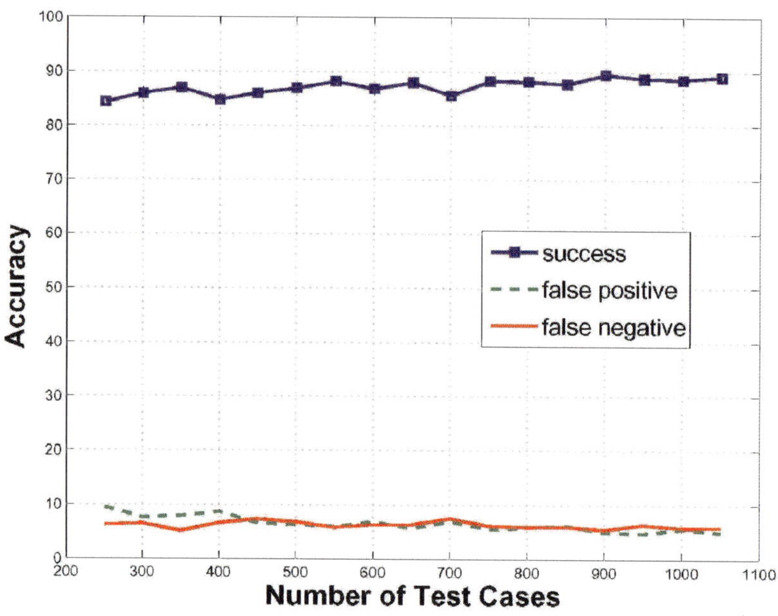

Figure 2.4. Results of the Automatic Identification of inputs for test sets with size $|T| = 250, 300, \ldots, 1000, 1050$. [18]

decrease in both the number of false positives and of false negatives) as the number of test cases increases from 250 to 1050. On average, this approach achieved an accuracy of 87% when test cases were randomly generated. To improve accuracy, the approach proposed in Section 2.2 is modified with the approach described in Section 2.3 and achieved an accuracy of 94%.

Figure 2.5 depicts the results of the simulation runs for the Adaptive SI Based approach as described in Figure 2.1. Similarly to the old approach, the accuracy does not present significant improvement as the number of test cases increases from 250 to 1050. However, when test cases were randomly generated, the improved approach achieved an accuracy of 94%, on average. This is 7% better than the approach reported in Figure 2.4. The addition of 20 test cases to the test set in every iteration where the eliminated inputs were not considered improved accuracy. The addition of test cases did, however, cause a significant decrease in false negatives (average of 0.76%) and a noticeable decrease in false positives (average of 5.50%). False positives are almost harmless because the controller controls an extra input

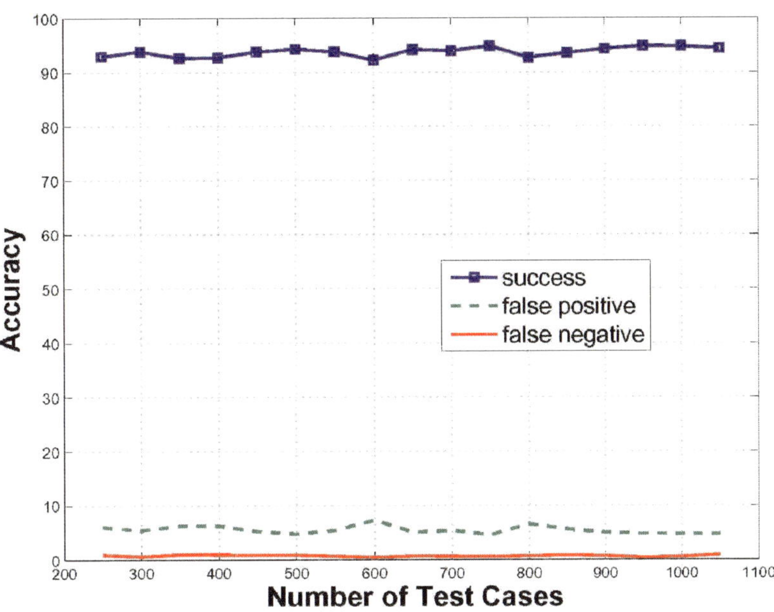

Figure 2.5. Results of the improved approach of the Automatic Identification of inputs for test sets with size $|T| = 250, 300, \ldots, 1000, 1050$.

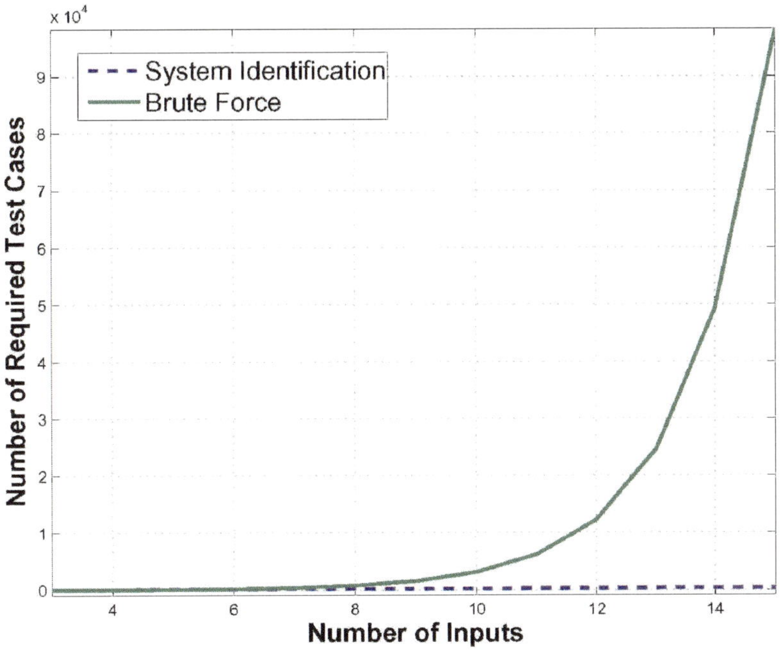

Figure 2.6. Comparison between the results of the propose approach using the system identification technique against the brute-force algorithm with $c = 3$.

that has no affect on the output. No more than one false positive has been identified per run in all the simulation runs. That is, the number of unnecessary test cases incurred by the extra input is small compared to a scenario where the number of false positives is greater than one. Unlike false positives, false negatives are harmful as the absence of the input(s) decreases the effectiveness of testing. However, the occurrence of false negatives is small (0.76%) and, similar to false positives, no more than one input is un-identified per simulation run.

Figure 2.5 shows that the Automatic Identification of inputs is not very sensitive to the test set size. The lower bound of $|T| = 250$ is due to limitations of the approach. That is, on average, the approach requires a minimum of 250 data points to compute the matrices' values from Eqn. 2.2. Using Eqn. 2.20 (assuming $c = 5$) leads to $|T| = 163840$ test cases needed for the brute-force algorithm. Comparing this with the results in Figure 2.5 shows an improvement of 156 times for the approach with only a small compromising of accuracy. The reduction of the number of test cases using the improved approach is substantial.

Table 2.1. Number of required test cases for the system identification and the brute-force approach with $c = 3$.

(a) Number of Inputs 3 - 9

Technique for Sensitive Input Identification	Number of Inputs						
	3	4	5	6	7	8	9
System Identification	115	120	145	160	180	205	210
Brute Force	24	48	96	192	384	768	1536

(b) Number of Inputs 10 - 15

Technique for Sensitive Input Identification	Number of Inputs					
	10	11	12	13	14	15
System Identification	220	240	255	280	295	310
Brute Force	3072	6144	12288	24576	49152	98304

Figure 2.6 shows that while the system identification approach presents a linear growth with respect to the number of test cases required for a given number of inputs, the brute-force algorithm presents an exponential growth. As seen in Table 2.1, the brute-force approach is more efficient for systems with fewer than 6 inputs but becomes extremely inefficient as the number increases. It is clear from Figure 2.6 and Table 2.1 that system identification outper-

forms the brute-force algorithm for systems with more than 5 inputs. Except for the static, source-code-based identification of resource-sensitive inputs (which does not depend on test cases but has restrictions with respect to the types of resources it can handle) [48],an approach with the same goal as this approach could not be found and, therefore, no comparison could be made.

2.5 Cut Off Value Analysis

As stated before, the cut-off variable z has an impact on the accuracy of the input identification procedure presented here. The goal is to maximize success rate while minimizing false negatives and false positives. Notice that it is more important to minimize false negatives than false positives. False negatives may lead to uncontrollability of the inputs and, consequently, failure to automatically achieve the desired level of stress or load for the particular resource. False positives may likely incur additional test cases, but this will not impact the results of Stress/Load Testing. To analyze the impact of the cut-off value 100 simulations runs have been executed for each value of $z = 0.5, 1, 1.5, \ldots, 5$. The average results are presented in Figure 2.7. Figure 2.7(a) shows that success rate reaches values around 94% for $z \geq 1.5$. Also, as can be seen in Figure 2.7(b), the smaller the value of z the smaller the percentage of false negatives. False positives have an opposite behavior. Values of z less than 1.5 lead to small number of false negatives, but false positives are too large in these cases. An intermediate solution is observed when $z = 1.5$. In this case, false negatives still present small values (0.64%) with reasonable values (5.4%) for false positives. Therefore, $z = 1.5$ seems to be the optimal value to achieve a reasonable success rate without compromising the number of false positives and false negatives.

2.6 Related Work: Feature Selection

Feature selection [35, 38, 25, 43, 39], also known as variable subset selection, is defined differently by many authors but all definitions are similar in intuition and/or content. Kira

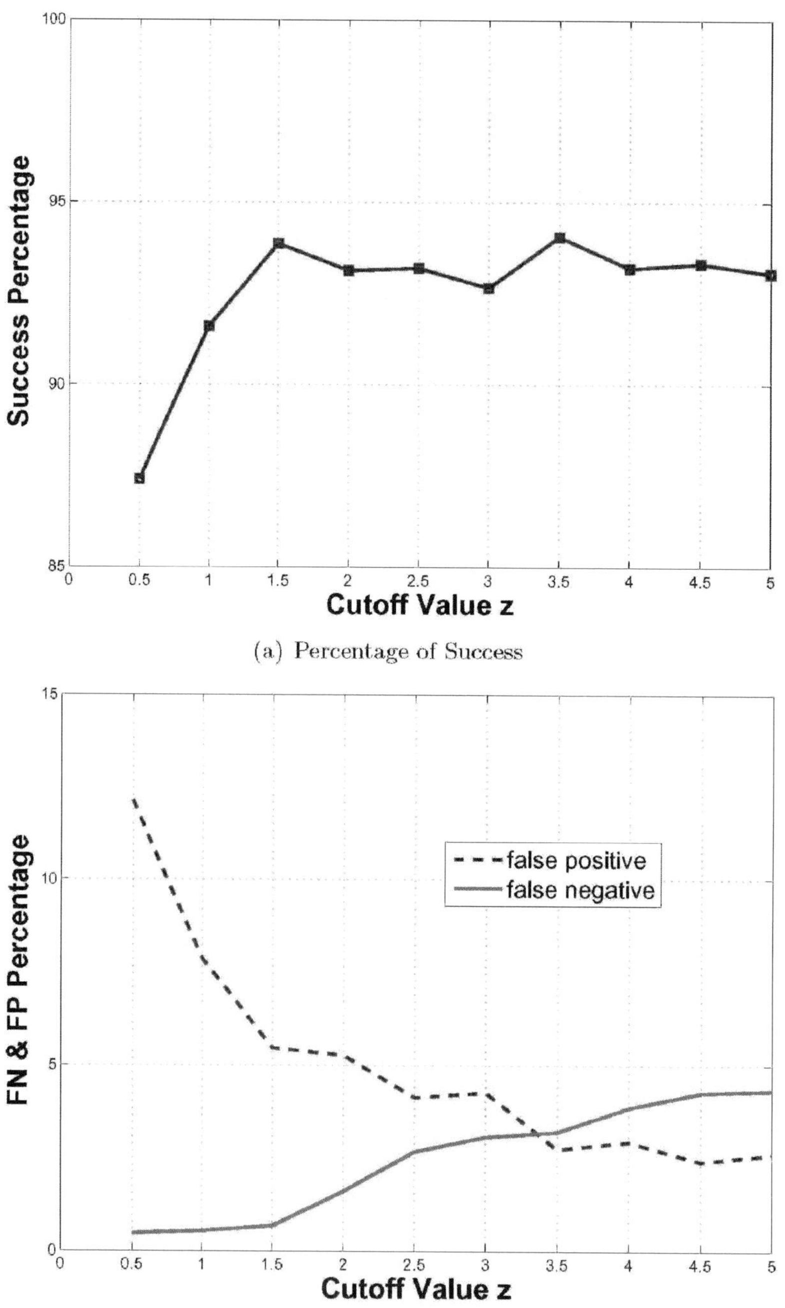

(a) Percentage of Success

(b) Percentage of False Positives (FP) and Negatives(FN)

Figure 2.7. Results for different cut off values.

and Rendell [38] defined feature selection as "the problem of choosing a small subset of features that ideally is necessary and sufficient to describe the target concept." Narendra and Fukunaga [43] described it as the problem "to select a subset of (m) features from a larger set of (n) features or measurements to optimize the value of a criterion over all subsets of the size m." Dash and Liu [25] described feature selection as an attempt to select the minimally sized subset of features according to a criteria where the criteria can be "the classification accuracy does not significantly decrease; and the resulting class distribution, given only the values for the selected features, is as close as possible to the original class distribution, given all features." Ideally, feature selection methods try to find the best subset among the competing 2^N candidate subsets of features according to some evaluation function where N is the feature set size.

Feature selection reduces the cost of recognition by reducing the number of features that need to be collected and in some cases it provides a better classification accuracy due to finite sample size effects. For example, if the task is to classify whether a document is about cats or not. The data provided is the word counts in the document and it is represented by the set X in the Table 2.2(a). Out of the whole set of words X, only three words "cat", "kitten", and "feline" can help determine whether the document is about cats or not and they are represented by the set $ReducedX$ in Table 2.2(b). Feature Selection has a variety of applications including Market Analysis, Customer knowledge, Quality control, Text Categorization, Machine vision, Bioinformatics, and System diagnosis.

A very important problem in feature selection is finding a feature subset that produces higher classification accuracy. For example, a physician may make a decision based on the selected features whether a dangerous surgery is necessary for treatment or not. For the purpose of this work, feature selection is used to identify inputs that are sensitive to a specific resource. For example, a system may have $\{I_1, I_2, \ldots, I_{15}\}$ as the set of inputs while only inputs $\{I_4, I_7, I_8, I_{11}\}$ significantly affect resource R_1.

Table 2.2. Feature selection Example

(a) X

X	
Word	Count
cat	2
and	35
it	20
kitten	8
electric	2
trouble	4
then	5
several	9
feline	2
while	4
. . .	
lemon	2

(b) Reduced X

$ReducedX$	
Word	Count
cat	2
kitten	8
feline	2

Different algorithms for feature selection were proposed. Figure 2.8 by Jain and Zongker [35] represents a taxonomy of available feature selection algorithms into broad categories. First, the algorithms are divided into those based on statistical pattern recognition (SPR) techniques, and those using artificial neural networks (ANN). The SPR category is then divided into those guaranteed to find the optimal solution and those that may result in a suboptimal feature set. The suboptimal methods are split into those that store just one "current" feature subset and make modifications to it, versus those that maintain a population of subsets. Final distinction is between algorithms that are deterministic, produces the same subset every time, and those that are stochastic which could produce different subsets on every run. Beneath each leaf node in the tree, sample feature selection algorithms are listed. Jain and Zongker [35] provide detailed descriptions for the above listed categories.

Figure 2.9 by Dash and Liu [25] represents a summary of the feature selection methods based on the 3 types of generation procedures. The complete generation procedures are divided into "exhaustive" and "non-exhaustive". In the "exhaustive" category, a method may evaluate all 2^N subsets, or it may perform a "breadthfirst" search to stop searching once an optimal subset is found. In the "non-exhaustive" category, there are different search

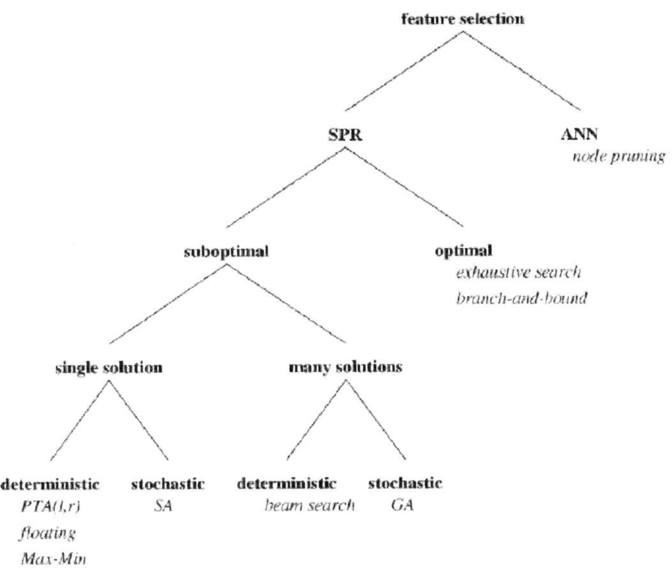

Figure 2.8. A taxonomy of feature selection algorithms

techniques such as "branch & bound", "best first", and "beam search". Heuristic generation procedures are divided into "forward selection", "backward selection", "combined forward/backward", and "instance-based" categories. Similarly, random generation procedures are divided into "type I" and "type II." "type I" and "type II" methods differ by the probability of a subset to be generated where it remains constant in "Type I" and changes as the program runs in "Type II". Dash and Liu [25] provide a comprehensive overview of the above listed feature selection methods.

A popular method of feature selection is sequential feature selection. It has two components. The first component is an objective function, called the criterion, used to select features and to determine when to stop. The second component is a sequential search algorithm which adds or removes features from a candidate subset while evaluating the criterion until there is no improvement in prediction. Sequential feature selection has two variants: Sequential forward selection (SFS) and Sequential backward selection (SBS). Sequential forward selection adds features sequentially to an empty candidate set until any further addition

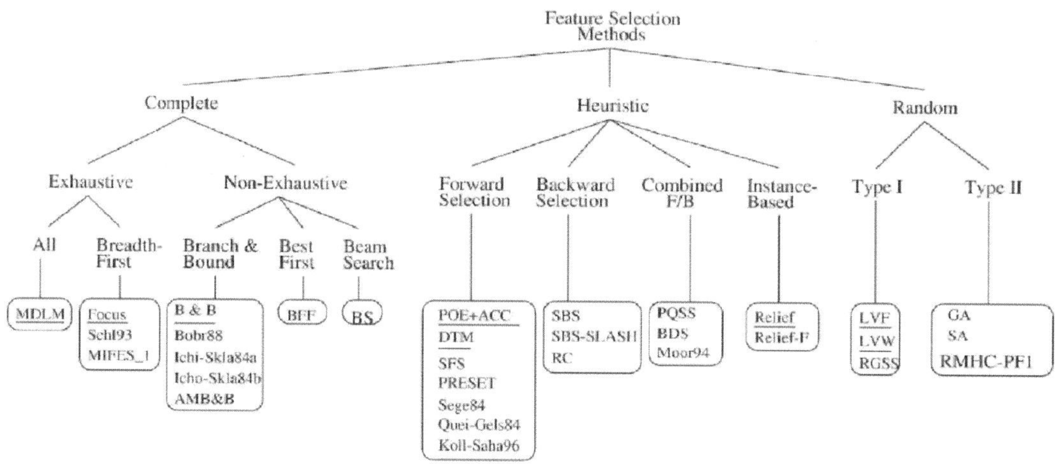

Figure 2.9. Summary of feature selection methods

does not decrease the criterion. Sequential backward selection removes features from a full candidate set until any further removal increases the criterion. To verify the experiments' applicability, the sequential feature selection algorithm available in MATLAB was chosen; and a comparison between the results of Feature Selection and Input Identification is presented in the next Section 2.7.

2.7 Input Identification vs. Feature Selection

As one of the most common used and powerful technique to identify sensitive input parameters Sequential Feature Selection is compared here with the proposed input identification. Before the results are compared it should be clear that the proposed approach goes beyond selecting inputs and produces a system that relates the inputs and outputs values. As described later in Section 4.2 this system is required to tune the PID controller. Furthermore, such system will also be required if multivariable control is considered.

Despite the functionality need for the proposed approach this section compares the results of applying both SFS and the Adaptive SI algorithm to randomly generated data. As presented in Section 2.4 five different functions $f_1, f_2, \ldots, f_5$ have been used to generate data. However, the first four function have a well established relationship between input and output and the results produced by both approaches were always above 90% of success rate. Function f_5 can produce a multitude of curves which makes the establishment of the input output relationship much harder and has been selected for the comparison.

Another aspect that impact the results is the tolerance value; that is, the threshold value to consider the input sensitive or not to the output. Since the approaches are significantly different the tolerances have been selected based on empirical results. In both cases the initial tolerance refers to the first value that produces reasonable results for false positives (as seen in Figure 2.10) and the last tolerance refers to a value that is still producing reasonable results for false negatives. The 10 tolerances values used for SFS were $[0.01, 0.025, 0.05, 0.075, 0.1, 0.25, 0.5, 0.75, 1, 1.25]$ while the 10 tolerance values for Adaptive SI algorithm were $[0.1, 0.25, 0.5, 0.75, 1, 1.25, 1.5, 1.75, 2, 2.25]$. These values correspond to the 10 x-axis values in Figure 2.10 below.

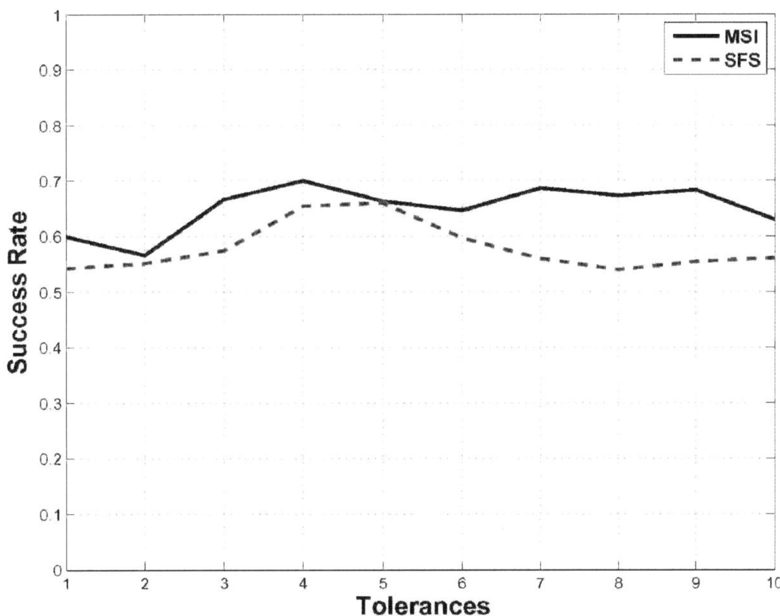

Figure 2.10. Feature Selection vs. Input Identification

As it can be seen in Figure 2.10 MSI has a slightly better success rate than SFS. The results in the figure are computed as the average of 100 simulations runs for each tolerance value. Therefore, in addition to provide a mathematical model that is required to tune the controller MSI also outperforms SFS when the relationship between inputs and outputs are not represented by standard mathematical functions.

2.8 Conclusion

In this chapter, a system-identification-based algorithm proposed to identify inputs that are sensitive to a specific resource. The proposed algorithm achieved an accuracy of 94% with simulation runs. A comparison with Feature Selection is done as well. Next chapter takes on another major part of the proposed approach, the controller.

CHAPTER 3
CONTROLLER

There are many types of control techniques available such as the PID Control, Direct Pole Placement, Adaptive Control, and Optimal Control among others. Though the "Controller" block in Figure 1.1 can be an instance of any of these controllers, a PID Controller has been used in the experiments discussed here. An instantiation of this controller is depicted in Figure 3.3. A PID Controller was chosen initially due to its simplicity, efficiency, and low overhead. An investigation of alternative control methods is deferred for now.

3.1 PID Controller

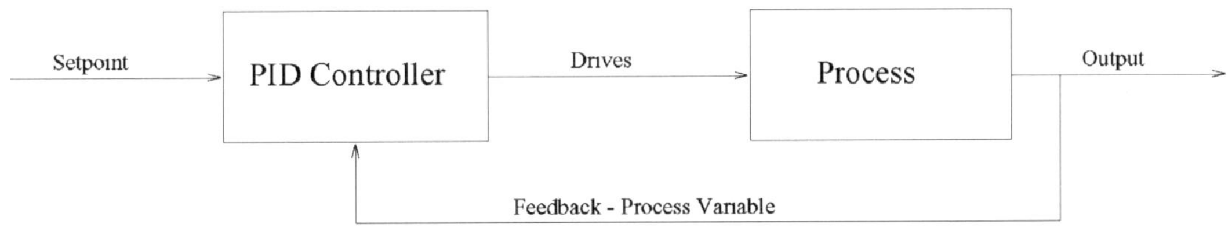

Figure 3.1. PID Controller

A PID Controller is often used in closed-loop control systems [30, 45, 44]. Closed-loop control systems utilize feedback that is a real-time measurement of the controlled process. The controlling device makes the *Process Variable* (the measured value of the process) follow the *Setpoint* (the desired value.) Usually, this requires tracking past values of both *Process Variable* and the *Setpoint*. Figure 3.1 represents a simple closed-loop control system. A full PID control algorithm with feedback is given in Eqn. 3.1.

$$y(t) = K_P e + K_I \int_0^t e dt + K_D \frac{de}{dt} \tag{3.1}$$

where y is the *Control Variable* (output) of the controller at time t, and e is the *Error* (error signal) at time t.

The proportional control ($K_P e$) deals with the current behavior of the process. The proportional control determines the Error, difference between the Setpoint and the Process Variable, and then applies appropriate proportional changes to the Control Variable to eliminate Error.

The integral control ($K_I \int_0^t edt$), focusing on the process's past behavior, adds long-term precision to a control loop. The integral control is needed to drive the process toward the Setpoint. It is the sum of all past *Errors* in the process over time.

The Derivative control ($K_D \frac{de}{dt}$) assists to reduce overshoot and improves response time. It monitors the rate of change of the Process Variable and make changes to the Control Variable accordingly. In other words, the Derivative control increases the process's stability by predicting process behavior.

K_P, K_I, and K_D, user-defined parameters, vary from one control system to another. These parameters must be tuned to optimize the precision of control. Determining these parameters' values is called PID Tuning. PID Tuning objectives are system stability, quick system response, minimum oscillations, and minimum Error value. Several methods are available for PID Tuning [30, 44].

Figure 3.2 presents an example of a PID Controller in a Cruise Control System. A Cruise Control System should control the speed of the vehicle by accelerating to the desired speed without overshooting and maintaining that speed with little deviation regardless of the weight of the car or the slope of the road. Cruise Controls are classic application of control system theory where the current speed of the vehicle is the *Process Variable* and the desired speed is the *Setpoint*. The Vehicle Speed Sensor monitors the vehicle speed and report it to the controller.

The proportional control adjusts the throttle proportional to the error where the error is the difference between the desired speed and the actual speed. The closer the car gets to the desired speed, the slower it accelerates. The integral of speed is distance and the integral control is based on the time integral of the vehicle speed error which is the difference between

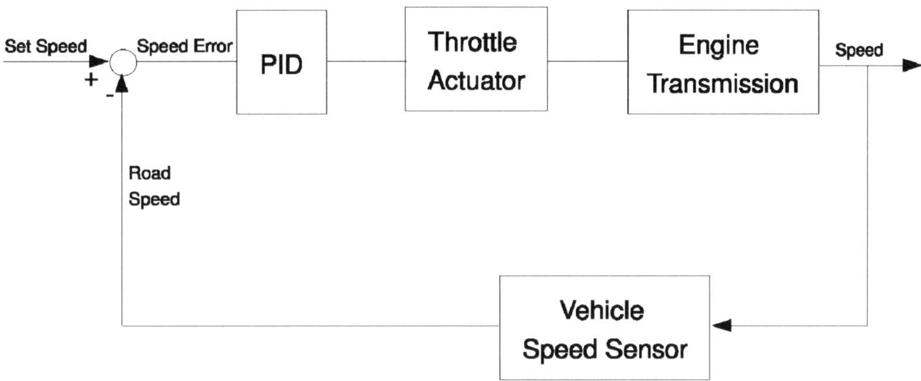

Figure 3.2. Cruise Control System.

the distance the vehicle actually traveled and the distance it would have traveled if it was going at the desired speed, calculated over a set period of time. The derivative of speed is acceleration and the derivative control helps the cruise control respond quickly to changes, such as hills.

PID controllers are most often used in continuous processes where the output is a continuous flow like a chemical process. However, PID control theory can also be used in a discrete manner and the next section describes how the proposed approach exploits the potential of using a simple control technique like a PID Controller to automate the Stress/Load testing process.

3.2 Using PID Controller in the proposed approach

A PID Controller is used to derive test cases automatically, and consequently, to achieve a pre-specified level of stress/load (the setpoint) for a specific resource. In Figure 3.3, the PID Controller accepts a setpoint as an input which represents the level of resource usage to be achieved by the application. The PID Controller uses an initial test case, defined by the tester, to drive the application to achieve that level of usage, independent of the initial

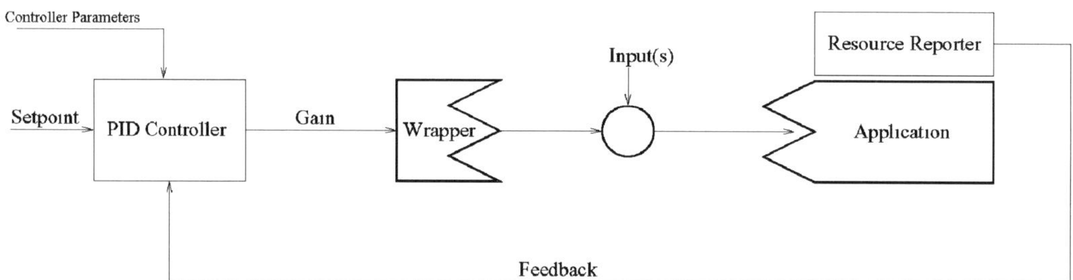

Figure 3.3. Instance of a controller used in Figure 1.1.

test case value. Every time the current usage is fed back to the PID Controller, the PID Controller computes a new gain. In general, positive gain represents an increase in input. Negative gain represents a decrease in input. To accurately represent the gain in terms of the input, wrappers (which are application specific) are used. Wrappers use the gain produced by the PID Controller and the previous set of input(s) to determine a new set of input(s). For example, assume the input represents the number of files to be compressed whose size should add to a specific value S. The wrapper will simply receive the new number of files and make the selection such that the summation of all individual sizes leads to S.

In addition to the application being tested, the approach requires a resource monitor which retrieves the current usage of the resource of interest and feeds this usage back to the PID Controller.

One important aspect to be considered for Stress and Load Testing is the simultaneous control of multiple resources [30, 41] such as memory, disk, bandwidth. The simultaneous control does not pose a problem in a scenario where the resource sensitive input sets are completely disjoint. That is, in Figure 1.1, if $\{R_1\ sensitive\ inputs\} \cap \{R_2\ sensitive\ inputs\} \cap \ldots \{R_n\ sensitive\ inputs\} = \emptyset$ then each controller can be applied independently. However, when there is an overlap among these inputs, a change in one input with respect to one resource may negatively impact another resource. In some cases, both resources may not be simultaneously controllable while, in other cases, overlapping inputs may be disregarded with minimum impact. The study of multi-variable control techniques is a subject of future work.

3.3 Advantages of Using a Controller

Two major issues prevent application of simple solutions such as the binary approach described in Section 3.4 to automation of the Stress and Load Testing process. These issues are the existence of warm-up periods and the presence of noise. The problems they cause and how the approach proposed here can properly handle them are described next.

Warm-up periods refer to the initial execution period where the software has not reached a stable behavior. For example, if the cache memory is empty, almost all pages will have to be loaded and, therefore, the response time will be different than when the cache memory is full and the likelihood of using a page from the cache is higher. Another example could be the dynamic allocation of memory (which could be dependent on some input parameters) during the start of the application. Warm-up periods may not be an issue if testers know how much time they should wait until start doing Stress/Load Testing. Results will be inconsistent if they start too early. The problem is that, in general, the warm-up period is unknown. A simple approach is to wait for a reasonably long period and, then, start testing. This is clearly not the best solution. Additionally, a new component may be launched later during the execution which could also have a warm-up period. The waiting time is not applicable in this case since the tester does not know when or if such component will be launched.

The second problem is the presence of noise (indeed, warm-up periods can also be considered as noisy periods). Noise consists of any element that perturbs the resource under consideration and the element is not under the control of the application being tested. For example, a server's response time depends not only on the time it takes to collect requested information and send it back to the client but also on ancillary tasks that the server may be executing. Ancillary tasks could consist of backup periods, disk cleaning or any other process that takes time off the main task. Continuous and abrupt changes in the amount of a noise within a system may result in an application with an almost ad-hoc behavior with respect to the resource to be stressed. In such a scenario, the application will be unstable. No technique can properly control the resource under consideration.

The solution proposed here for these two problems offers a major advantage of using an approach based on control theory. Noise has a common place in physical system. Controllers, such as the PID used here, have been designed to minimize and react to the effect of noise. For example, a cruise control continuously adjusts to road changes such as when the vehicle starts up or down on a hilly road. The same applies to software. The Zip experiment (presented in Section 5.3) is a good example; as the operating system starts other tasks, response time changes and the controllers react to these changes to keep the resource of interest within control limits. The same is true for warm-up periods. Setpoint may be achieved initially during the warm-up period, but as resource usage changes during the remaining warm-up period, a controller adjusts the inputs properly to compensate. As mentioned above, as long as the unexpected changes are neither too constant nor too large, the controller can properly compensate for those changes. Without a feedback control mechanism it is unlikely that automatic test-case generation for Stress and Load Testing will achieve reasonable success. Using an existing control mechanism is proposed here; the initial experiments' results indicate clearly the applicability and accuracy of such mechanisms.

3.4 Comparison with the Binary Approach

There are many Stress and Load testing tools in the industry. These tools require the interaction of the Software Tester for every run of the system (every test). To achieve a certain level of Stress Test, the tester provides inputs to stress test the application. Based on each run's results, the tester determines the next run's inputs. One could claim that this process could be automated using a binary approach to determine the next set of input values to use.

In a binary approach, if the current result is less than the desired setpoint, the inputs are doubled. If the current result is more than the desired setpoint, the current input is added to the last input which resulted in an output less than the desired setpoint and divide the sum by 2.

Figure 3.4 represents the results of using the binary approach to achieve a specified response time. As shown in the graph, a binary approach could not achieve reasonable level of stability. Oscillations around the setpoint were too large. At some points, the Average Response Time was almost double the desired setpoint. This clearly indicates that a binary approach does not provide an ideal algorithm to automatically control and drive the Stress Testing process. A binary approach could be applicable if the resource under control presents a well-defined behavior and no noise is observed in the process. This is rarely the case and, therefore, use of such approach would lead to inconsistence results.

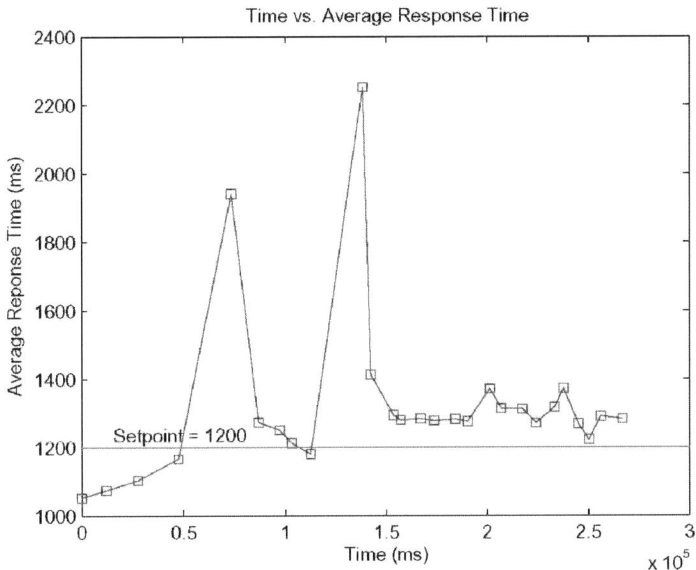

Figure 3.4. Application of Binary Approach to generate test cases to achieve a pre-defined level of response time for a client server system.

3.5 Conclusion

The proposed approach uses control theory as it has proven successful in a wide range of applications. Also, it has been designed to minimize and react to the effect of noise and warm-up periods. More specifically, a PID Controller was chosen due to its simplicity, efficiency, and low overhead. The PID controller internal parameters (K_P, K_I, and K_D)

need to be chosen, knows as PID Tuning, and a software tester is not assumed to have the expertise for that. Therefore, tuning a PID controller is presented in the next chapter.

CHAPTER 4
AUTOMATIC CONTROLLER TUNING

Once the inputs affecting a resource have been identified, a controller can be used to automatically change the inputs and to drive the resource to a specified level of stress/load. However, this can be achieved in many ways. For example, one may want to achieve a load level as early as possible. However, this may cause oscillation unacceptable for load testing. If the load level is crossed, Stress has been performed rather than Load Testing. In another scenario, oscillation around the set point may be acceptable. Consider a web-server where the number of requests that can be handled within a certain response time is to be measured. In this case, the controller can oscillate around the set point as long as it reaches stability later.

Internal parameters of controllers are used to achieve control under one of the scenarios described above. For example, the PID Controller described in Section 3.2 has three parameters K_P, K_I, and K_D. A large gain (determined by a large value for K_P) will, most likely, lead to oscillations. K_D can then be used to decrease the oscillations while K_I drives the resource to the set point with the smallest error possible–leading to a more stable and accurate solution.

Testers are not assumed to be familiar with control theory and to be able to define the controller internal parameters' values on their own. However, testers are assumed to be able to define which scenarios are more appropriate for testing an application while the controller's parameters should be computed automatically according to the testers' specifications. Computation of a controller's parameters is known as tuning [30]. Tuning Techniques such as Lambda Tuning [14], Cohen-Coon [24], and Ziegler-Nichols [50] allow automatic tuning of a controller.

Figure 4.1 shows the system response based on three different tuning set ups based on a setpoint change from 0 to 1.

- Critically Damped: The system shown in curve (a) is optimally tuned and its said to be critically damped. It does not overshoot the setpoint value and settles as quickly as possible (6 seconds) at the new setpoint value without oscillating.
- Over Damped: The system shown in curve (b) is over damped. It does not overshoot the setpoint; however, it takes too long to reach the desired set-point without oscillating.
- Under Damped: The system shown in curve (c) is under damped. It overshoots the setpoint and then oscillates around the setpoint.

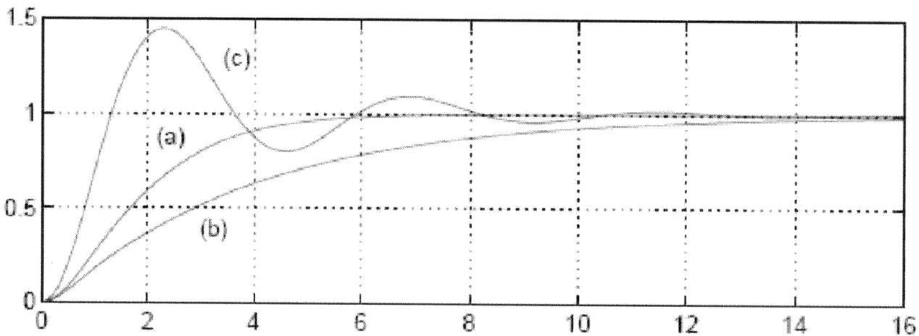

Figure 4.1. Typical Pid Tuning Graph

From a software testing perspective, in most cases, Critically Damped or Over Damped curves are preferred. In general, it does not matter how soon to reach a load level or a stress level as long as the system achieves that load and stays there. So, Critically Damped or Over Damped curves are desirable in software testing. On the other hand, in some scenarios, Under Damped curves must be avoided. For example, if load testing is performed, the system should not pass a certain a load (setpoint) and therefore the system should not be under damped.

4.1 Ziegler-Nichols Tuning Method

Ziegler and Nichols [50] conducted numerous experiments and proposed rules for determining values of K_P, K_I, and K_D based on the transient step response of a plant. They proposed several methods, two of these methods are discussed here for the purpose of this work. Closed Loop Tuning is discussed in Section 4.1.1 and Open Loop Tuning is discussed in Section 4.1.1.

Controller Tuning is a very important phase of the overall controller design and it critically determines the performance of the control loop which in turn affects the overall quality of the product. Control performance directly affects product quality whether its a chemical process control, a steel plant, or a software application as in this work. The basic objective is to meet the product's specifications and limitations. Depending on the application, different performance criteria has to be chosen. Rather than stating the specifications of the control loop as a complete reference model, some behavioral parameters of the reference model are often used like Overshoot, Rise Time, Settling Time, Decay Ratio, and Steady-state error. Figure 4.2 describes the control performance criteria (Overshoot, Rise Time, Settling Time, Decay Ratio, and Steady-state error). The Rise Time is the amount of time it takes for the system to initially cross the setpoint. Overshoot is the maximum amount that the response goes over the setpoint and it is represented by a (height of the first peak) in the Figure 4.2. Settling Time is the time required to get the response with a +/-.05% band of the setpoint. In other words, the time it takes the response to meet its goal. This time should be as small as possible because smaller values represent a faster response. Decay Ratio is the ratio of the heights of successive peaks in the response and it is represented by b/a (ratio of the second overshoot by the first overshoot) in the Figure 4.2. Steady-state Error is the difference between the steady-state output and the desired output. For example, in the step response of the closed loop model, the overshoot can be stated to be lower than a certain value (for example less than 10 percent), the Rise Time can be specified to be less than so many seconds or minutes which means the control loop will quickly rise to the setpoint, the settling time can be stated to be less than so many seconds or minutes which means after the settling time the error is almost zero, or the decay ratio has to be of certain amount which is the ratio of the second overshoot by the first overshoot. If the decay ratio is small, it means even if there is a first overshoot which is slightly large, it quickly comes down; in other words, the error is rapidly converging to zero. Table 4.1 summarizes the effects of increasing each of the controller parameters K_P, K_I, or K_D on the Control Loop behavioral parameters.

Table 4.1. Effects of increasing a control Parameter (K_P, K_I, or K_D) independently on the Control Loop behavioral parameters

Response	Rise Time	Overshoot	Settling Time	S-S Error
K_P	Decrease	Increase	Small change	Decrease
K_I	Decrease	Increase	Increase	Eliminate
K_D	Minor Decrease	Decrease	Decrease	No effect

Stability is a consideration and where gain margins and phase margins come into existence. Phase and Gain margins are basically protections that the loop will not oscillate even if the process loop changes by a little bit or the overall loop phase changes by a little bit.

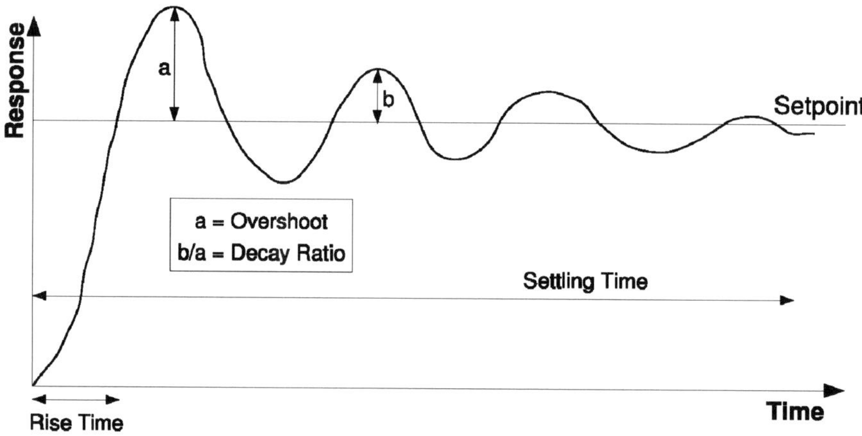

Figure 4.2. Pid Controller Common Performance Criteria - Overshoot, Rise Time, and Settling Time

4.1.1 Ziegler-Nichols: Closed Loop Tuning Method

One of the methods that Ziegler and Nichols [50] proposed is the Closed Loop tuning method. Closed Loop procedure is essentially an iterative procedure which means it involves trial and error. First, the loop need to be in a stable operating condition. Second, the actions of the integral and derivative modes need to be minimized and if its necessary the proportional gain need to be reduced as well to get the loop in a stable state. Third, make a step change in the setpoint. The process loop rises and then dies down; if the cycles damps out like in Figure 4.3, increase the gain which means the process will be close to an oscillation. Keep on increasing the gain till the loop oscillates like in Figure 4.4. Once the loop oscillates, it will provide two values: the ultimate gain G_u and the ultimate period P_u (oscillation period) as shown in Figure 4.4 where the ultimate gain is the gain which made the plant oscillates. The Close Loop Ziegler-Nichols tuning method uses the variables P_u and G_u to generate the controller parameters as described in Table 4.2. For Close Loop Tuning, some trial and error is needed and in most cases the setting obtained are to be treated as good first estimates and further trial and error may be needed.

Table 4.2. Ziegler-Nichols Tuning Method: Closed Loop

Controller Type	Proportional Gain (K_P)	Integral Gain (K_I)	Derivative Gain (K_D)
P	$0.5G_u$	-	-
PI	$0.45G_u$	$1.2K_P/P_u$	-
PID	$0.6G_u$	$2K_P/P_u$	$K_P P_u/8$

4.1.2 Ziegler-Nichols: Open Loop Tuning Method (Reaction Curve)

Another controller tuning method proposed by Ziegler and Nichols [50] is based on open loop step response. A very popular way of open loop experimentation is the Reaction Curve where the step response experiment is applied to the open loop plant so the output rises resembling an S-shaped curve with no overshoot. This S-shaped curve is called the reaction curve as shown in Figure 4.5. First, the process must be "lined out" which means the response is not changing. Second, change the output by a small amount. Third, the step response is going to

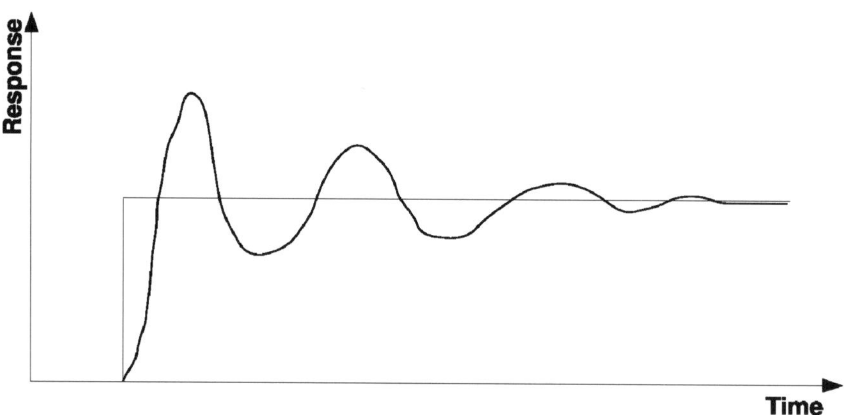

Figure 4.3. Closed Loop Tuning - After increasing the proportional gain, the process loop rises and then dies down

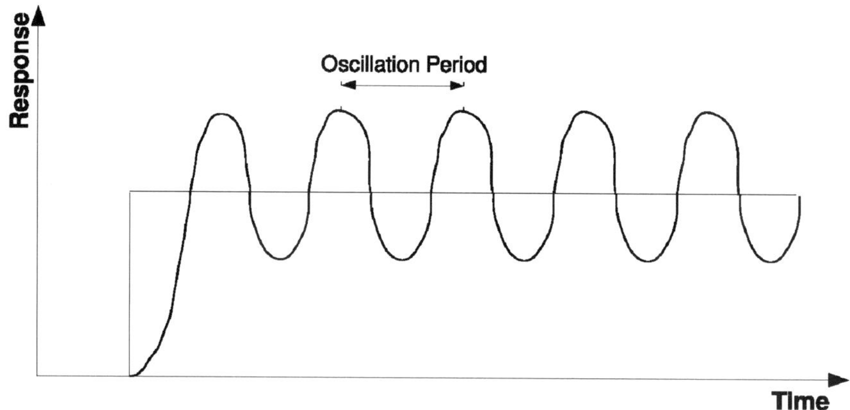

Figure 4.4. Closed Loop Tuning - After increasing the proportional gain, the process loop oscillates

rise similar to the curve in Figure 4.5. X represents the percentage of change in the response values, R is the slope of the tangent line at the Point of Inflection (POI), and L is the time it takes the curve to rise (point of intersection between the tangent line and the x-axis). The Open Loop (Reactive Curve) Ziegler-Nichols tuning method uses the variables X, R, and L to generate the controller parameters as described in Table 4.3. The Open Loop (Reactive Curve) Ziegler-Nichols tuning method only needs a single experiment as no trial and error is needed like in the Close Loop tuning method described in Section 4.1.1. However, for some processes loop can not be opened and determining the slope can be difficult with the presence of noise. Another limitation is that it applies only to over damped processes.

Table 4.3. Ziegler-Nichols Tuning Method: Open Loop Reaction Curve

Controller Type	Proportional Gain (K_P)	Integral Gain (K_I)	Derivative Gain (K_D)
P	X/LR	-	-
PI	$0.9X/LR$	$0.3K_P/L$	-
PID	$1.2X/LR$	$0.5K_P/L$	$0.5K_PL$

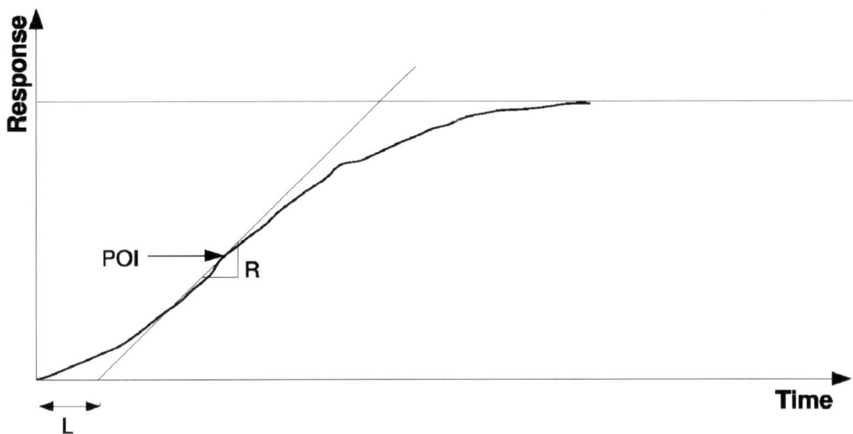

Figure 4.5. Open Loop Tuning - Reaction Curve

4.2 Automatic Controller Tuning for Software Applications

For software application systems, in most cases, its not feasible to apply the closed Loop tuning method described in Section 4.1.1. Unlike industrial control applications (regulation of speed, temperature, flow, pressure and other process variables), its not possible to achieve the oscillations with fixed oscillation period needed for the tuning since most software systems are stable and not asymptotically stable [41]. For example, in the Web Application Experiment described in Section 5.4, if there are 100 clients sending requests to the JBoss Server, these 100 clients will report slightly different average response time each time depending on the network and other tasks that the server machine and the client machines are performing.

In Chapter 2, System Identification Technique was used to identify the subset of inputs that affect the output (the resource of interest). As observed in Section 2.3, The adaptive algorithm generated a model that represents the software system under test. The generated model is used to automatically tune the PID controller to efficiently control the software system as described below.

The Controller Tuning for software applications is done in two steps. First, the value of the K_P is determined. Second, once the K_P value is known, the K_I and the K_D values are calculated. Figure 4.6 presents the pseudo-code for the PID tuning algorithm. In this algorithm, in addition to the system model $sysModel$ and the set of sensitive inputs $SensitiveInputs$ generated in Section 2.3, the tester needs to provide the desired level of stress/load to achieve (*setpoint*), an initial test case ($initialTestCase$), the minimum number of test cases required before reaching the *setpoint* ($minSteps$), and the maximum number of test cases to execute before reaching the *setpoint* ($maxSteps$).

The function "simulateData" represented in Figure 4.8 generates data using $sysModel$, $SensitiveInputs$, *setpoint*, K_P, and $initialTestCase$. It uses the command $lsim$ available in MATLAB to simulate the response of the model to arbitrary inputs. First, it simulates the response of the model ($sysModel$) to the initial test case ($initialTestCase$) where *position* represented the response reported back. Then, it applies the proportional Gain K_Pe to the current set of input to generate a new set of input where $K_Pe = K_P \times error$ and $error = setpoint - position$. Once the new input data set is generated, after applying the proportional gain, the model is simulated again and a new *position* is reported. The algorithm keeps on doing so until the *position* reported is greater than or equal to the *setpoint*.

Figure 4.6 represents the pseudo-code for the PID Tuning algorithm. Using only the proportional control (K_Pe), initially the integral control ($K_I \int_0^t edt$) and the Derivative control ($K_D \frac{de}{dt}$) are set to 0, the algorithm tries to achieve the desired *setpoint* starting with an initial K_P value of 0.0002. It invokes the function "simulateData" using $K_P = 0.0002$ and if the size of the input set generated is less than *minSteps*, it decreases the K_P ($K_P = K_P/1.1$) and tries again. However, if the size of the input set is greater than *maxSteps*, it increases the K_P ($K_P = K_P * 1.1$) and tries again. It modifies the value of K_P until the size of the input set, test cases needed to reach the *setpoint*, is within *minSteps* and *maxSteps* ($minSteps < size(Inputs) < maxSteps$). Once that is achieved, the control parameter K_P is determined.

Next, the values of K_I and K_D are determined using the Open Loop Tuning Method described in Section 4.1.2. Using the size of the input data set and the response generated in the previous step, nonlinear regression function available in MATLAB is used to estimate the coefficients of the function defined in Figure 4.7. Then, the tangent line at the point of inflection is determined. Finally, using K_P and L, the values of K_I and K_D are determined based on the formulas of Open-Loop Tuning described in Table 4.3 where $L = x0$ and $x0$ represents the intersection of the tangent line with the x-axis.

4.3 Conclusion

A PID tuning algorithm, based on Ziegler-Nichols tuning technique, applicable to software applications is proposed here. Once the inputs affecting a resource have been identified and

```
function PidTuning(sysModel, setpoint, initialTestCase, minSteps, maxSteps,
                                                              SensitiveInputs)
  kp = 0.0002;
  global setpoint; // global variable setpoint
  [Output, steps] = simulateData(sysModel, setpoint, kp, initialTestCase,
                                                              SensitiveInputs)
  while(size(steps) < minSteps || size(steps) > maxSteps)
    if(size(steps) < minSteps)
      kp = kp / 1.1
    elseif(size(steps) > maxSteps)
      kp = kp * 1.1
    end
    [Output, steps] = simulateData(sysModel, setpoint, kp, initialTestCase,
                                                              SensitiveInputs)
  end

  //Non Linear regression
  syms x;          //x is the symbolic variable
  t=0:1:size(steps)
  BETA0=[5, (6/max(t))];

  try
    beta = nlinfit(t,Output,@expFunction,BETA0)
  catch
    beta = BETA0
  end

  fit = expFunction(beta,x)
  fit_dt1 = diff(fit)              // First Derivative
  fit_dt2 = diff(fit_dt1)          // Second Derivative
  //Inflection Point: set second derivative to 0
  xInflectionPoint = solve(fit_dt2)
  yInflectionPoint = subs(fit, x, xInflectionPoint)
  //Slope of the Tangent Line
  slope = subs(fit_dt1, x, xInflectionPoint)
  //Equation of the Tangent Line
  tangentLine = slope*(x-xInflectionPoint)+yInflectionPoint
  //Intersection of the Tangent Line with the x-axis
  x0 = ((0-yInflectionPoint)/slope)+xInflectionPoint

  L = abs(x0)       //L is the absolute value of x0
  constant = 0.6;
  ki = (constant/L) * kp
  kd = (constant * L * kp)
end
```

Figure 4.6. Pseudo-code for the Automatic Tuning algorithm

```
function Y = expFunction(beta,x)
  b = beta(1);
  k = beta(2);
  global setpoint; // global variable setpoint
  Y  = setpoint*exp(-b*exp(-k*x))
end
```

Figure 4.7. Pseudo-code for the function representing the exponential model.

```
function [y, t] = simulateData(sysModel, setpoint, kp, initialTestCase,
                                                               SensitiveInputs)
  position = 0;

  //simulate model response to the initialTestCase using lsim()
  [position, Inputs] = simulateDataUsingLsim(sysModel, initialTestCase,
                                                               SensitiveInputs)

  while (position < setpoint)
    error = setpoint - position
    gain = kp * error

    Inputs = ApplyProportionalControl(Inputs, gain, SensitiveInputs)

    //simulate model responses to Inputs using lsim()
    [position, Inputs] = simulateDataUsingLsim(sysModel, Inputs,
                                                               SensitiveInputs)
  end
end
```

Figure 4.8. Pseudo-code for the simulateData function

the controller has been tuned, the controller can be used to automatically change the inputs and to drive the resource to a specified level of stress/load. Next chapter presents several experiments to prove the applicability and accuracy of the proposed work.

CHAPTER 5
EXPERIMENTAL RESULTS

Four experiments are discussed in this section. The first experiment α, examined control of the amount of memory an application used. The experiment properly automated the load testing process and showed that a controller can change the inputs properly to drive memory usage to a desired level. Also, memory leaks can be detected using a controller and the allocated memory's behavior. β, the second experiment, introduces interesting aspects such as the control of more than one variable and a less deterministic behavior of the resource under control–response time in this case. In the third experiment, γ, the application under test is the Linux utility Zip and the resource under control is the response time. The fourth experiment, δ, is the most realistic example so far. The application under test is a Web Application hosted by a JBoss Server. Also, in δ, the resource under control is the response time.

The approach controlled memory usage in α, response time in γ where response time is the amount of time it takes to compress a number of files into an archive, and response time in β and δ where response time is the amount of time it takes a web server to handle a client request. Controlling different resources proves the flexibility of the proposed approach. Using Controller Tuning in Experiments γ and δ allowed to bring the system under test rapidly to the desired load and maintain that load for several time slots which attested the efficiency and accuracy of the approach. Using the approach with different types of applications, simple application (written in C language) in α, remote client-server application in β, Linux utility Zip in γ, and a web application hosted by a JBoss Server in δ showed that the approach is widely applicable.

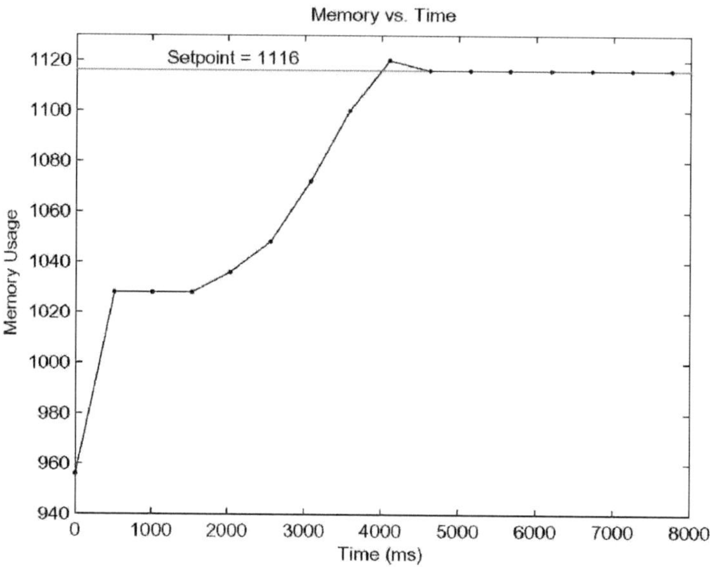

Figure 5.1. Results of applying a PID Controller to achieve a specified level of memory usage without oscillations.

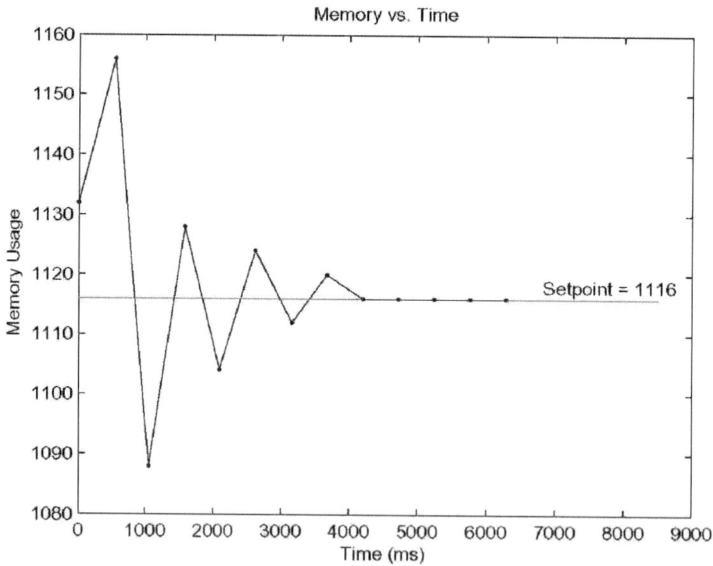

Figure 5.2. Results of applying a PID Controller to achieve a specified level of memory usage with oscillations.

5.1 Experiment α: Memory Usage

Experiment α is indeed a toy-example that uses two-dimensional data multiplication. The goal was to show the approach's applicability. The application under test is a matrix multiplication which accepts one input n of type integer. The application randomly creates two matrices $A_{n \times n}$ and $B_{n \times n}$, multiplies the two matrices, and stores the result in a third matrix $C_{n \times n}$. The resource of interest is memory; the resource monitor feeds back the matrix application's memory usage to the PID Controller after processing a new test case.

The wrapper accepts the gain and the last executed test case as inputs to generate a new test case. For example, if the last executed test case had an order $n = 10$ and the PID Controller produced a gain of 50%, then, the wrapper generates a new test case where the order is $n = 15$. However, if the PID Controller produced a gain of -30%, then, the new test case would have been $n = 7$. These values are also dependent on the other control parameters K_I and K_D.

Figure 5.1 shows the results of starting the application with an initial test case $n = 2$ and a desired memory usage of 1116 (setpoint). The figure shows that the initial matrix order is 2 with a memory usage of 956. As time passes, the PID Controller produces positive gains, and the matrices' size continues to increase until memory usage exceeds the setpoint to 1120. At this point, the matrices' size coalesces achieving the desired setpoint of 1116 for memory usage. In Figure 5.1, notice the small overshoot, where the x-axis represents the time in milliseconds, the y-axis represents the memory usage by the matrix application so far, and the horizontal line represents the desired memory usage. The overshoot could be minimized or canceled by tuning the PID Controller, more specifically by changing the K_P value.

Figure 5.2 shows a different run of the system with an initial order input of 120 and a setpoint of 1116. After 26 ms, the application created the matrices, multiplied them, and allocated a total memory of 1132. The memory usage oscillated around the setpoint before coalescing at the desired level of 1116. Figure 5.2 shows the changes in memory usage with respect to time, and it clearly shows the oscillations of the memory usage. PID Tuning (briefly described in Chapter 4) plays a major role in oscillations. In other words, the values of K_I, K_P, and K_D can be tuned to minimize or even eliminate oscillations.

In addition to achieving Stress and Load Testing, the approach assists in detecting continuous memory leak. While the resource usage increased, the PID Controller produced positive gains and the produced test cases changed in direction X. After reaching the setpoint, if the resource usage continued increasing, PID Controller produced negative gains and the produced test cases changed in a way opposite to direction X. This behavior indicates a memory leak. In the matrix application example, when the memory usage passed the setpoint and kept increasing, the PID Controller produced a negative gain which resulted in decreasing the order of the matrix until it reached an order of zero. Exceeding the setpoint in memory usage, at the same time, the order of the matrix was decreasing revealed the injected memory leak.

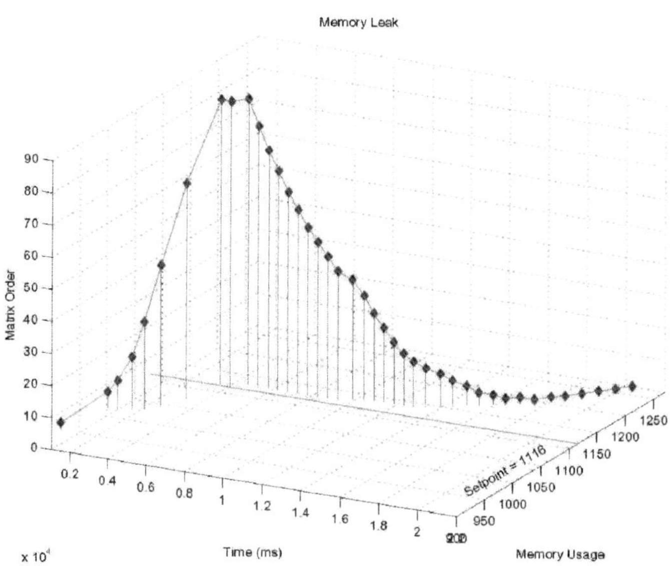

Figure 5.3. Results of applying a PID Controller to potentially identify memory leaks.

Figure 5.3 shows the results of an example where a memory leak has been injected into the matrix application. The setpoint provided by the tester was 1116. The PID Controller drove the memory usage up to the desired level by continuously increasing the Matrix order up to 87. After reaching the desired level, the memory continued to increase beyond the setpoint as a result of the memory leak. To accommodate that increase, the PID Controller produced negative gains, thereby, decreasing the matrix order. The continuous increase in memory usage eventually resulted in an order size of 0 for the matrices.

5.2 Experiment β: Response Time

Nowadays, web applications are the most widely used software applications. An overloaded Web site is a major problem and, once happening, the problem often cannot be addressed rapidly. To avoid overloading the application in production, Load and Stress Testing becomes essential. Testing a web application involves testing techniques such as Load Tests, Stress Tests, Failover Tests, Soak Tests, Performance Tests, and Network Sensitivity Tests. However, one of the most fundamental tests is a Load Test which is an end-to-end performance test under an anticipated production load. The test's primary objectives are to determine the response times for time-critical transactions and to measure the application's ability to function correctly under load. Response time has been defined in two ways that are appropriate to discuss here: "... the interval between the receipt of the end of transmission of an inquiry message and the beginning of the transmission of a response message to the station originating the inquiry. [27]"; "The elapsed time between the end of an inquiry or demand on a computer system and the beginning of a response. [34]" For example, response time may be the time between an indication of the end of an inquiry and the display of the first character of the response at a user terminal. However, the first definition of response time is not usually appropriate within a performance-related, application requirement specification. The first character returned to the application often does not contribute to the rendering of the screen with the anticipated response. For the work discussed here, the response time is defined as the interval between the receipt of the end of transmission of an inquiry message and the beginning of the transmission of a response message. Slow response time distracts users and makes it more annoying and difficult to interact with a site. With this in mind, load testing the server becomes more important. The number of clients the server can handle without exceeding a pre-determined response time must be determined.

In this experiment, a remote client-server application is considered (emulating a web server). Response time was the resource of interest. The system consisted of a Server that accepted client requests, processed the requests, and sent responses to the clients. The goal was to load test the server by adding clients until a specified response time (Setpoint) was achieved. In other words, if the Setpoint was 5ms, the goal was to determine how many clients the server could handle without exceeding the 5ms response time.

There were two types of clients in the experiment, Regular Clients and VIP Clients. The server served VIP Clients faster than Regular Clients. On average, the server took twice as much time to process a Regular Client than a VIP Client. In this experiment each client (both VIP and Regular) sent requests to the server continuously. After sending a request and receiving a response, the server saved the response time in a log file. After 5 consecutive request/response cycles, the average response time was computed and reported to the controller. When all the existing clients had reported their average response times, the controller calculated the gain and either added new clients or deleted existing clients.

Two variables affected the Average Response Time: (i) Total number of clients in the system (totalClients); and (ii) the percentage of regular clients in the system (perRegClient). Average Response Time increased as the number of Clients increased. Also, the Average Response Time increased as the percentage of Regular Clients increased because Regular Clients took more time to be processed. Results achieved when controlling one variable are discussed in Section 5.2.1, results achieved when controlling one variable of two inputs are discussed in Section 5.2.2, and results achieved when controlling two variables of two inputs are discussed in Section 5.2.3.

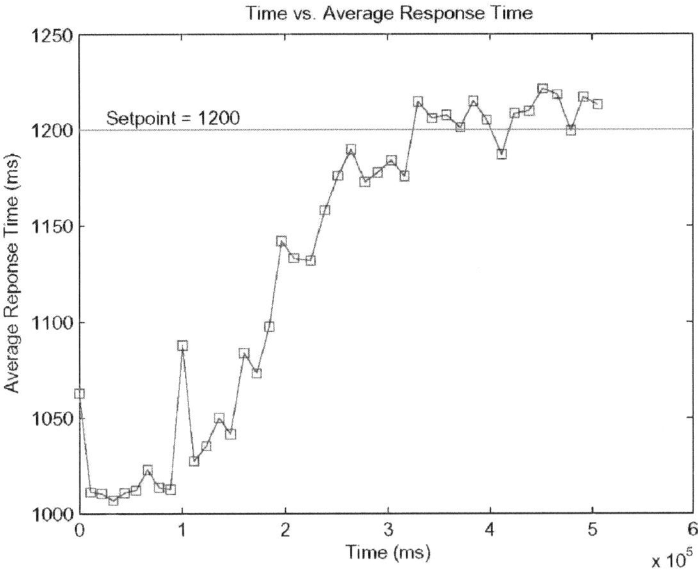

Figure 5.4. Application of a PID Controller to achieve a certain average response time for the remote client server application. Only regular clients are considered and the total number of clients is the only input variable used by the controller.

5.2.1 Controlling one variable

In this first experiment, only Regular Clients were considered (the server processed all clients equally). Because the system only had Regular Clients, the perRegClient variable was always 100%. Therefore, totalClients (total number of clients) was the only variable to control.

As described earlier, when all the existing clients have reported their average response time to the controller, the controller calculates the gain. The wrapper accepts the gain produced by the PID Controller and the last executed test case (number of currently running clients) to generate a new test case. For example, if 10 clients are currently running and the PID Controller produces a gain of 50%, then, the wrapper adds 5 clients. However, if the PID Controller produces a gain of -30%, then, the controller keeps 7 clients and deletes 3 clients.

In Figure 5.4, the x-axis represents the time in milliseconds, the y-axis represents the Average Response Time in milliseconds of all clients currently running, and the horizontal

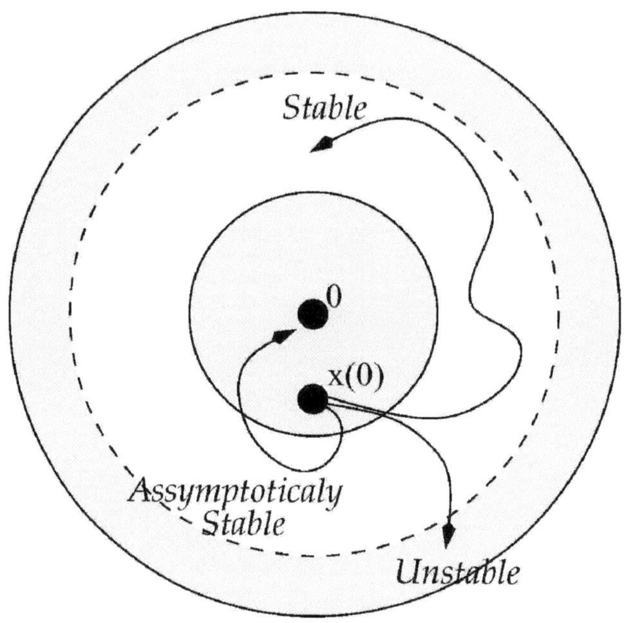

Figure 5.5. Stability Graph.

line represents the desired average response time. The system started with an initial test case of totalClients=5 and the desired Average Response Time was specified to be 1200ms (setpoint). As time advanced and the PID Controller produced positive gains, the number of clients increased until the Average Response Time reached the setpoint of 1200ms. It required 18 clients to reach a setpoint of 1200ms. Unlike the memory usage experiment, response time for a remote client server application is less deterministic. That is, while memory usage can be completely controlled and, in most cases, driven to the exactly specified level and remain there (this represents an asymptotically stable system [41] as shown in Figure 5.5), the same does not occur for response time. The server, though dedicated, still has to handle OS-related requests or backups which affects the response time. That is, the average response time will oscillate around a value, but it will not always be the same value. Figure 5.4 clearly shows this behavior. After creating 18 clients, the average response time oscillates around the setpoint (this represents a stable system [41]). Stability properties of a system can be analyzed when a model is available. The model derived using the system identification technique for input identification (Chapter 2) can also be used for this purpose. However, stability analysis of the system is deferred to future work.

In this experiment, only one variable (totalClients) was used to control the system and achieve the desired average response time. However, most applications have several inputs affecting the resource. Next, the experiments presented have two variables affecting the Average Response Time (resource). The results of these experiments clearly indicate the necessity of properly identifying and controlling inputs that significantly affect the resource to be stressed.

5.2.2 Controlling one variable of two inputs

In this experiment, both Regular and VIP clients are considered. As described earlier, the server served VIP Clients faster than Regular Clients. Therefore, not only the total number of clients but also the percentage of Regular Clients affected the Average Response Time. To demonstrate the need to identify and control all the inputs affecting the resource, only the variable totalClients is used to reach the setpoint while perRegClients changed randomly. That is, after all the existing clients reported their average response time to the controller, the wrapper accepted the gain produced by the PID Controller and the total number of currently running clients to determine the new number of clients. However, the percentage of perRegClients was generated randomly.

The system in Figure 5.6 started with 2 clients (totalClients) and with perRegClients set to 50% (1 Regular Client). As time advanced, the total number of clients increased due to the gains produced by the PID; but the number of Regular and VIP clients changed randomly. This behavior can be observed in Figure 5.7. Because only one of the variables that affected the resource was under control, the behavior of the Average Response Time could not achieve a reasonable level of stability, and oscillations around the setpoint were too large (Figure 5.6). This behavior clearly indicated that both variables need to be controlled. One could argue that one variable could be controlled while all others are kept constant; however, this would significantly decrease the effectiveness of a Test Suite. Keeping part of

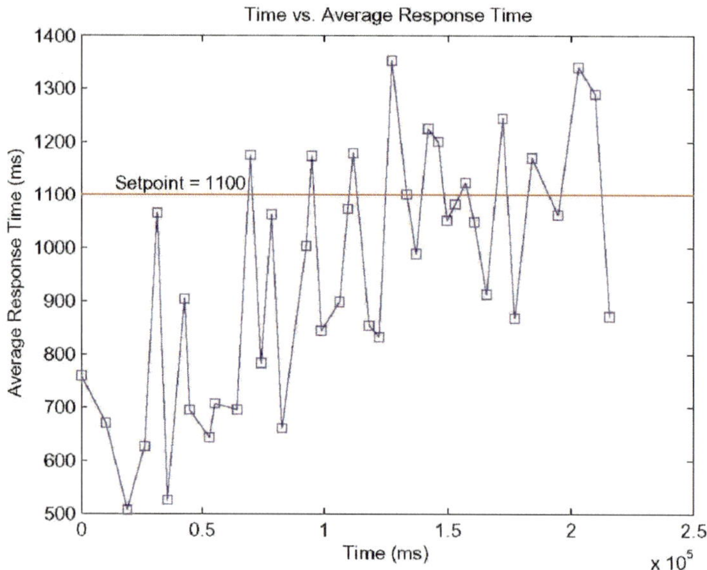

Figure 5.6. Application of a PID Controller to achieve a certain average response time for the remote client server application. Both Regular and VIP clients are considered and the total number of clients is the only input variable used by the controller.

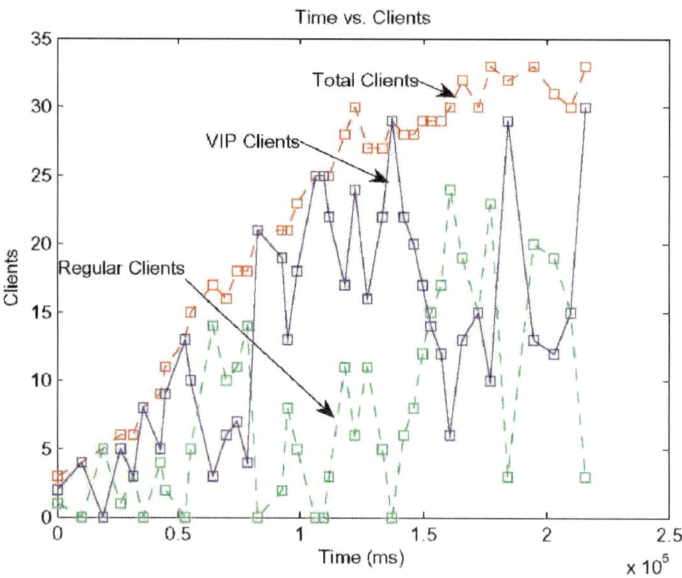

Figure 5.7. Total number of clients and number of Regular and VIP clients used in Figure 5.6. The percentage of regular clients is randomly determined.

the inputs constant would lead to many important combinations of relevant input variables not being included in the Test Suite and as a result not being tested.

5.2.3 Controlling two variables of two inputs

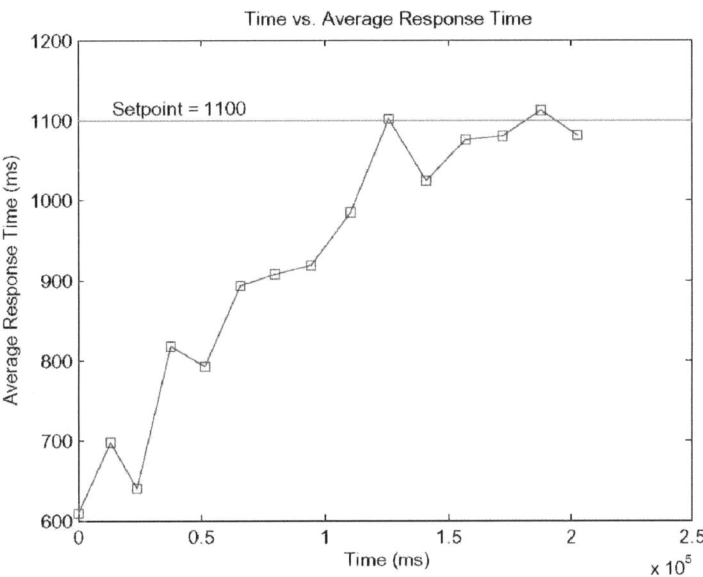

Figure 5.8. Application of a PID Controller to achieve a certain average response time for the remote client server application. Both Regular and VIP clients are considered; total number of clients and percentage of Regular Clients are both used by the controller.

The difference between this experiment and the one described in the previous section is that the two variables totalClients and perRegClients are controlled to achieve the Setpoint.

The system in Figure 5.8 started with 15 clients (totalClients) with perRegClients set to 13.33% (2 Regular Clients). After the existing clients had reported their Average Response Time to the controller, the controller calculated the gain. The wrapper accepted the gain produced by the PID Controller as an input in addition to totalClients and perRegClients. The wrapper determined the new number of clients (totalClients) and the new percentage of Regular Clients (perRegClients).

For example, if 10 clients were running with 40% regular Clients (4 Reg and 6 VIP) and the PID Controller produced a gain of 50%, then, the wrapper would add 5 clients to

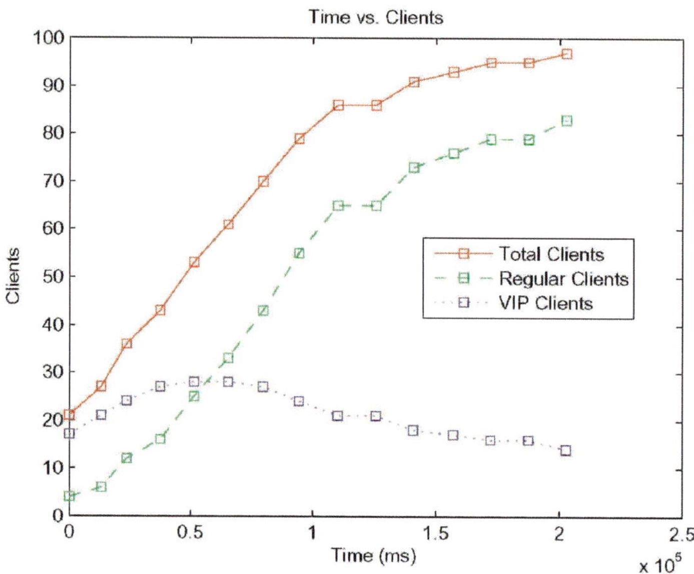

Figure 5.9. Total number of clients and number of Regular and VIP clients used in Figure 5.8. The percentage of regular clients is determined by the controller.

set totalClients to 15 and increase perRegClients to 60%. So, the system would end with 9 Regular Clients and 6 VIP Clients. If the PID Controller produced a gain of -30% then the controller would keep only 7 clients and decrease perRegClients to 12%. So, the system would have one Regular Client and 6 VIP Clients. As can be seen in Figure 5.8, when both variables affecting response time were controlled, the system achieved the desired Average Response Time. When compared with results reported in Figure 5.6 (1 controlled variable), the oscillations around the setpoint were considerably smaller. Figure 5.9 shows that the number of clients changed in a more systematic manner when both variables were controlled.

5.3 Experiment γ: Zip Experiment

Experiment γ used the Unix utility Zip[1] to apply the approach to a widely used application to test its suitability with a real application. First, random data (files) were collected as

[1] "Zip is a compression and file packaging utility for Unix, VMS, ... The program is useful for packaging a set of files for distribution; for archiving files; and for saving disk space by temporarily compressing unused files or directories." [12]

inputs. The input set had four file types: PDF, TXT, JPG, and PS. File sizes ranged from 1 KB to 10000 KB. Second, the Zip application was executed 250 times. In each execution, 8 inputs were considered and each input's value was randomly selected. Table 5.1 briefly describes each input.

Table 5.1. Inputs to the Zip Application.

Input	Description
fileTypes	Type of files to be zipped. Four different types of files were considered PDF, TXT, JPG, and PS. The combinations of file types were decided randomly.
Size	Total size of the files to be compressed. Its a number between 1 and 10000.
numFiles	Number of files to be zipped. Its a number between 1 and 100.
compRatio	(-#) Regulate the speed of compression using the specified digit #, where -0 indicates no compression (store all files), -1 indicates the fastest compression method (less compression) and -9 indicates the slowest compression method (optimal compression, ignores the suffix list). The default compression level is -6.
Integ	(-T) Test the integrity of the new zip file. If the check fails, the old zip file is unchanged and (with the -m option) no input files are removed. This input has two values either True or False.
storePath	(-j) Store just the name of a saved file (junk the path), and do not store directory names. By default, zip will store the full path (relative to the current path). This input has two values either True or False.
LMF	(-o) Set the "last modified" time of the zip archive to the latest (oldest) "last modified" time found among the entries in the zip archive. This can be used without any other operations, if desired. This input has two values either True or False.
Encrypt	(-e) Encrypt the contents of the zip archive using a password which is entered on the terminal in response to a prompt (this will not be echoed; if standard error is not a tty, zip will exit with an error). The password prompt is repeated to save the user from typing errors. This input has two values either True or False.

At the end of each run of the system, the inputs values and the execution time were logged. Execution time was the time it took the Zip application to compress the input files into an archive. Next, the Input Identification Approach (Chapter 2) was applied to the test data which represented the matrix *Matrix_Results* in the Input Identification approach. The cut-off value was 1.5%. Using the Input Identification Approach, inputs that were sensitive to the resource of interest (execution time in this case) were identified. Table 5.2 depicts the results of applying the approach. The Sensitivity Percentage represents the likelihood that an input affects the resource. An input with a Flag of 1 affects the resource; otherwise, a Flag of 0 means the input has no noticeable effect on the resource.

5.3.1 Identifying Sensitive Inputs

The System identification Procedure described in Section 2.3 is used to identify the sensitive inputs. Initially, all inputs are considered to affect the resource and, hence, all flags are set to 1. In the first iteration the approach eliminated the input "Encrypted" because its sensitivity value was less than 1.5. Therefore, its flag was set to 0. In the second iteration, two additional inputs were eliminated: "storePath" and "Integrity". In the third iteration, the input "LMF" was eliminated. The following three iterations did not affect the set of sensitive inputs. Finally, the set of sensitive inputs was comprised of "Size", "numFiles", and of "CompRatio." At first analysis, only "Size" and "CompRatio" could be selected as inputs affecting the response time. However, the response time is not only the CPU time but the entire time the application takes to return the compressed files. Therefore, the number of files impacts considerably on response time. More files will lead to more I/O interruptions and, consequently, a longer time to complete the compression process. Using System Identification Procedure produced accurate results in this case.

5.3.2 Tuning the PID Controller

After determining the sensitive inputs (size, compRatio, and numFiles), the controller has to be tuned using the algorithm described in Section 4.2. The goal is to achieve a *setpoint* of 500 ms where *minSteps* is set to 10 and *maxSteps* is set to 30. Initially, using the system model generated by the Input Identification Approach (Chapter 2) and the proportional gain (K_Pe), the algorithm tries to achieve the *setpoint* of 500 ms. Starting with $K_P = 0.0002$, the algorithm took more than 30 test cases to reach the *setpoint*. It kept on changing the

Table 5.2. Results of Applying the System identification Procedure to the Zip application random test cases.

Iterations		Inputs							
		Size	num-Files	comp-Ratio	Integ	store-Path	LMF	Encrypted	file-Types
1	Sensitivity	56.78	9.02	16.87	2.95	3.34	2.31	1.05	7.68
	Flag	1	1	1	1	1	1	0	1
2	Sensitivity	61.67	7.15	18.06	1.17	0.25	4.62	1.19	5.88
	Flag	1	1	1	0	0	1	0	1
3	Sensitivity	60.41	10.22	19.20	4.07	0.53	1.46	2.97	1.13
	Flag	1	1	1	0	0	0	0	0
4	Sensitivity	52.48	14.37	14.61	0.22	4.33	0.18	1.61	12.19
	Flag	1	1	1	0	0	0	0	0
5	Sensitivity	64.52	9.49	16.28	4.59	0.35	0.41	2.91	1.46
	Flag	1	1	1	0	0	0	0	0
6	Sensitivity	60.00	10.77	20.61	3.18	0.22	0.87	3.49	0.85
	Flag	1	1	1	0	0	0	0	0

K_P until it reached a value of 0.057062 and the system was able to reach the *setpoint* with 29 steps. The results are depicted in Figure 5.10.

Next, non-linear regression is used to model the curve in Figure 5.10 as an exponential model which is represented in Figure 5.11. Using the curve in Figure 5.11, a tangent line is drawn at the Point of Inflection (POI). Using the time delay $x0$ and the generated K_P, the values of K_I and K_D are determined to be 0.012809 and 0.040673 respectively.

5.3.3 Controlling the Zip Application

After determining the sensitive inputs (size, compRatio, and numFiles) and tuning the controller, the controller controlled the Zip application and drove its execution time to the desired stress level by controlling the sensitive inputs. Figure 5.12 presents the results of controlling 3 inputs where the setpoint is 500 ms. After reaching the setpoint, the controller maintained the output within $\pm 1\sigma$ (standard deviation) for 205 points (71%), $\pm 2\sigma$ accounts for 64 points (23%), and $\pm 3\sigma$ accounts for 19 points (6%). This indicates that the system is stable but not asymptotically stable as it oscillates always within the $\pm 3\sigma$ range. Notice that the $\pm 3\sigma$ range is the standard used to determine if a process is stable for statistical

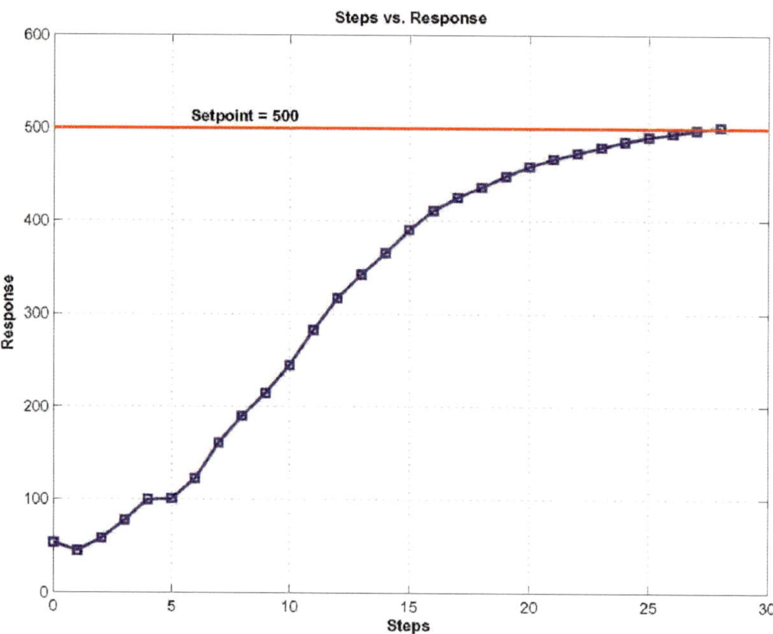

Figure 5.10. Tuning the controller for the Zip Experiment - Determining K_P

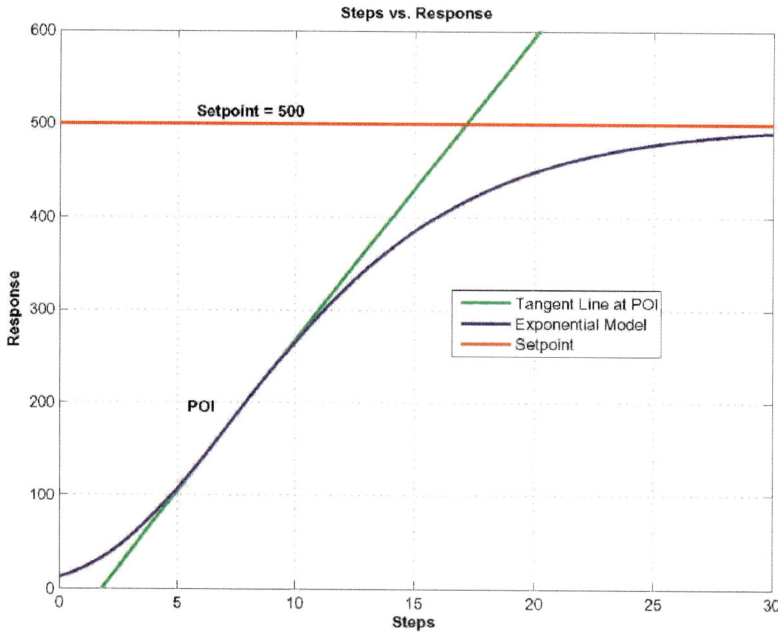

Figure 5.11. Tuning the controller for the Zip Experiment - Determining K_I and K_D

process control [46, 28].

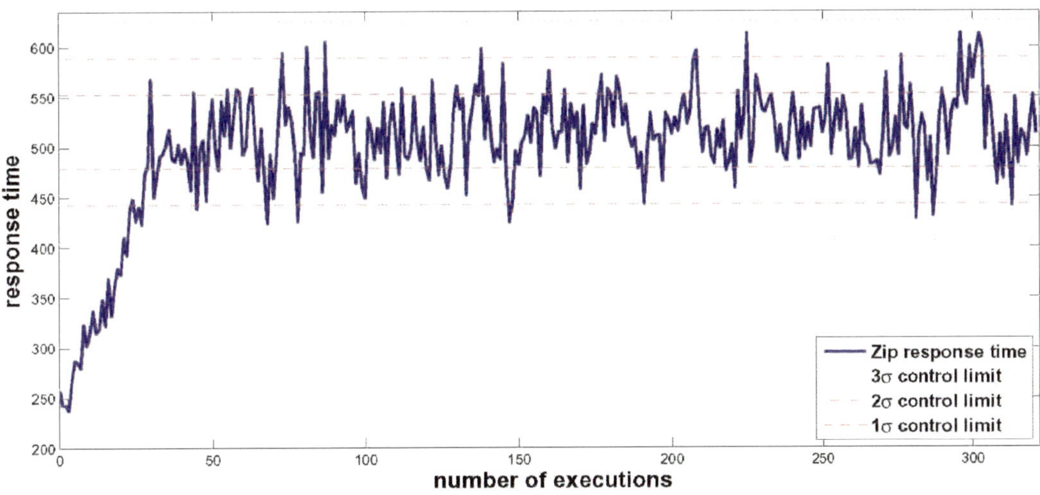

Figure 5.12. Results of experiment γ where the three input variables "Size", "numFiles", and "CompRatio" are used to control response time.

5.4 Experiment δ: Web Application

Experiment δ used a Web Application Server hosted by a JBoss Server to apply the approach to the most widely used software applications. Response time was the resource of interest. Just like in the Experiment β, the response time here is defined as the interval between the receipt of the end of transmission of an inquiry message and the beginning of the transmission of a response message.

The Experiment hardware setup is represented in Figure 5.13. There are 5 Client Machines, a Server Machine, and a Network Switch. The Server Machine hosts the PID Controller and the JBoss Web Server where the Web Application is hosted. The Network Switch joins the Client Machines and the Server Machine together within one local area network (LAN). Each Client Machine hosts a number of Java Clients that invoke the Web Service hosted by the Web Application on the Server Machine.

Figure 5.14 represents a Software high level overview of the Experiment. The system consisted of a JBoss Server that accepted client requests, processed the requests, and sent

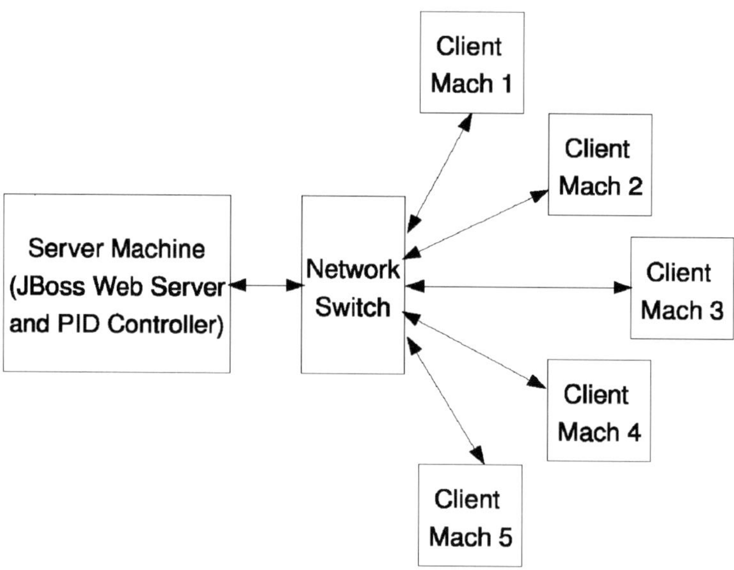

Figure 5.13. Web Application Hardware Setup

responses to the clients. The JBoss applications server is a J2EE platform for developing and deploying enterprise Java applications, web applications and services, and portals. The JBoss applications server hosts the BookStoreApp web application which exposes the BookStore Web Service. The goal was to load test the BookStoreApp web application by adding clients until a specified response time (Setpoint) was achieved. In other words, if the Setpoint was 5000ms, the goal was to determine how many clients the server could handle without exceeding the 5000ms response time. The HttpClientController performs the action of killing/creating HttpClient processes.

Each HttpClient sent SOAP requests to the server continuously. After sending a SOAP request and receiving a SOAP response, the server saved the response time locally. After 5 consecutive SOAP request/response cycles, the average response time was computed and saved in a Buffer. When all the existing HttpClients had reported their average response times, the ResourceMonitor sent the average response time to the PID Controller. The PID

Controller calculated the gain and passed it to the HttpClientController who either added new clients or deleted existing clients.

HttpClient can be of different types where the type is defined based on the amount of data returned by the BookStore web service as part of the SOAP response. There are 7 different types of HttpClients where Type1 receives one data record in the SOAP response, Type10 receives 10 data records, Type25 receives 25 data records, Type50 receives 50 data records, Type100 receives 100 data records, Type500 receives 500 data records, and Type1000 receives 1000 data records.

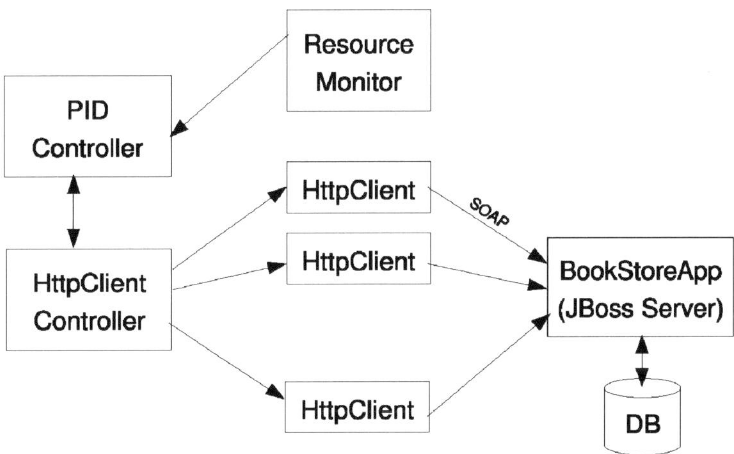

Figure 5.14. Web Application Software Setup

Figure 5.15 presents the communication between the BookStore web service and the Java Client HttpClient. Web Service Description Language (WSDL) is an XML-formatted language used to describe the semantics and interfaces of the web service BookStore. HttpClient uses the WSDL to create the necessary code to invoke the web service and to receive a response back using SOAP messages.

Two things affected the Average Response Time: (i) Total number of clients in the system (totalClients); and (ii) the percentage of each type of client in the system. Average Response Time increased as the number of Clients increased. Also, the Average Response

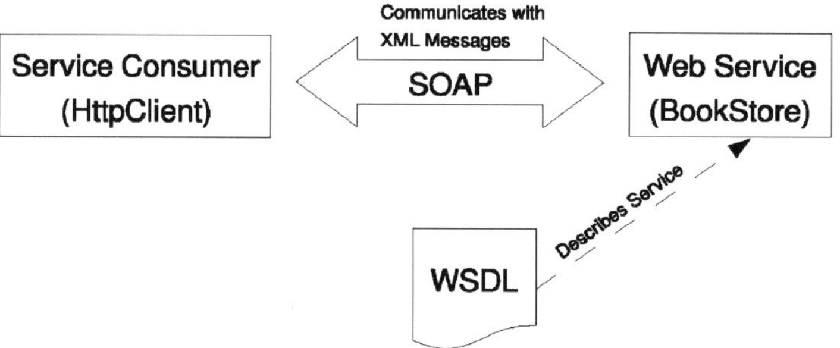

Figure 5.15. Web Service Overview

Time increased as the percentage of higher types of clients increased. The results of tuning the PID Controller for this experiment are presented in Section 5.4.1. Results achieved when controlling one type of client are discussed in Section 5.4.2, controlling two types of clients in Section 5.4.3, controlling three types of clients in Section 5.4.4, controlling four types of clients in Section 5.4.5, and controlling seven types of clients in Section 5.4.6.

5.4.1 Tuning the PID Controller

To tune the PID controller for the Web Application, using the algorithm described in Section 4.2, the goal is to achieve a *setpoint* of 5000 ms where $minSteps = 10$ and $maxSteps = 50$. Using the PID Tuning algorithm, the system reached the *setpoint* of 5000 ms in 49 steps with $K_P = 0.003221$. The results are depicted in Figure 5.16.

Figure 5.17 represents the exponential model of the curve in Figure 5.16. Using the Figure 5.17, a tangent line is drawn at the Point of Inflection (POI). Using the time delay $x0$ and the generated K_P, the values of K_I and K_D are determined to be 0.000195 and 0.008525 respectively.

5.4.2 Controlling One Type of Client

Table 5.3 represents the system setup for the first run of the experiment. The system started with 5 total clients of Type1. Type1 clients made 100% of the total number of clients so

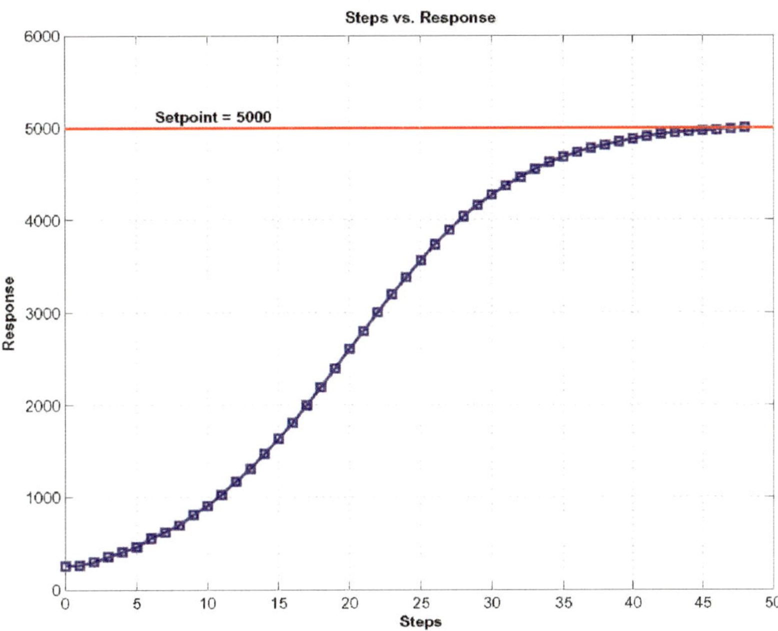

Figure 5.16. Tuning the controller for the Web Application Experiment - Determining K_P

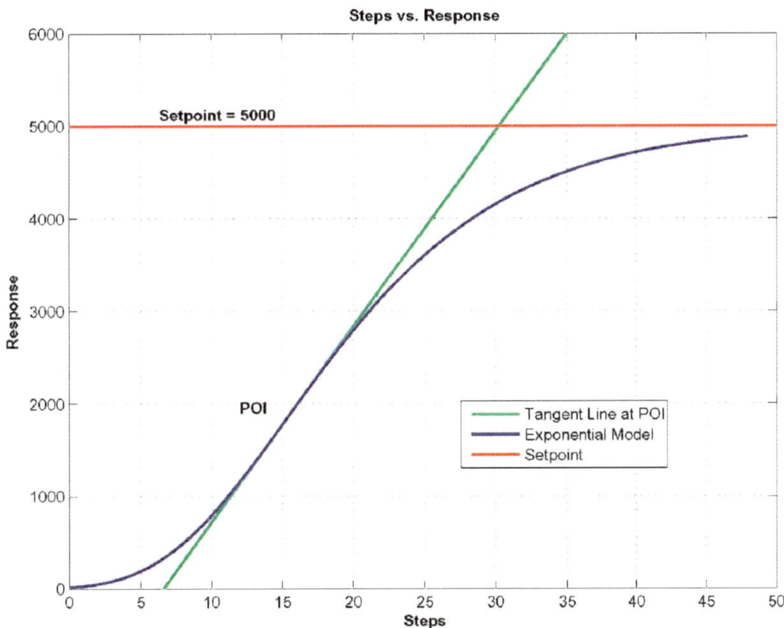

Figure 5.17. Tuning the controller for the Web Application Experiment - Determining K_I and K_D

they were the only clients to control.

Table 5.3. Web Application Setup - Controlling 1 type of clients.

Client Type	Percentage	Initial Test Case
Type 1	100	5
Type 10	0	0
Type 25	0	0
Type 50	0	0
Type 100	0	0
Type 500	0	0
Type 1000	0	0

As described earlier, when all the existing clients have reported their average response time to the controller, the controller calculates the gain. The HttpClientController uses the gain produced by the PID Controller and the last executed test case (number of currently running clients) to generate a new test case and either adds new Type1 clients or drop existing ones. For example, if 10 clients are currently running and the PID Controller produces a gain of 50%, then, the HttpClientController adds 5 clients. However, if the PID Controller produces a gain of -30%, then, the controller keeps 7 clients and deletes 3 clients.

In Figure 5.18, the x-axis represents the time in milliseconds, the y-axis represents the Average Response Time in milliseconds of all the clients currently running, and the horizontal line represents the desired average response time. The system started with an initial test case of totalClients=5 and the desired Average Response Time was specified to be 5000ms (setpoint). As time advanced and the PID Controller produced positive gains, the number of clients increased until the Average Response Time reached the setpoint of 5000ms. After reaching the setpoint, the average response time oscillates around the setpoint (this represents a stable system [41]).

In Figure 5.19, the x-axis represents the time in milliseconds and the y-axis represents the total number of clients running in the system. It required 97 clients to reach a setpoint of 5000ms. After creating 97 clients, the number clients needed to maintain the 5000ms setpoint oscillated around 95 clients.

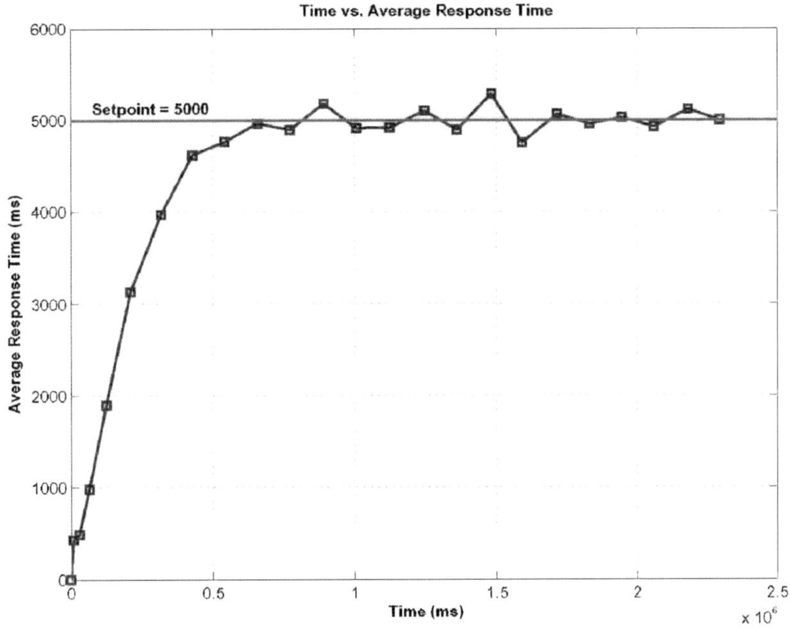

Figure 5.18. Results of controlling one type of client (Type1) to achieve a 5000 ms response time.

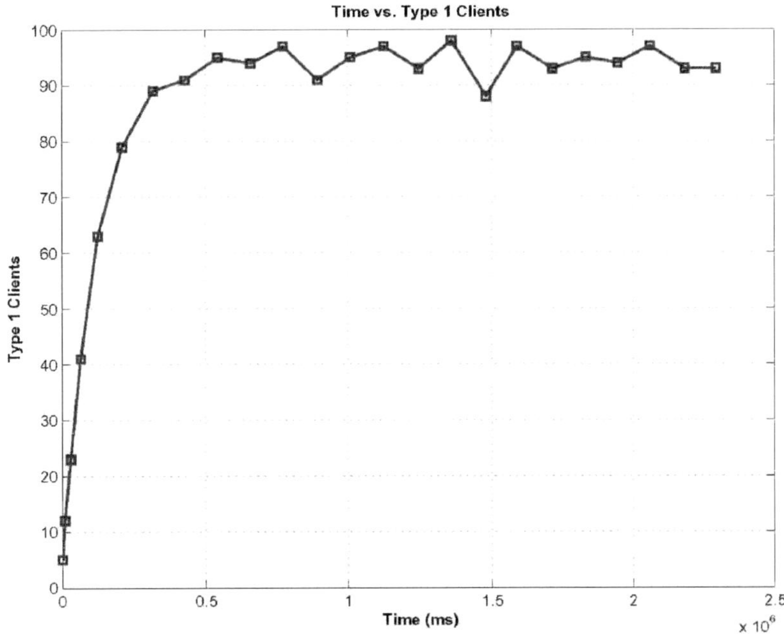

Figure 5.19. Total number of Type1 clients used in Figure 5.18.

5.4.3 Controlling two Types of Clients

Table 5.4 represents the system setup for the second run of the experiment where 2 types of clients were controlled. The system started with a total of 10 clients, 5 of Type1 and 5 of Type50. Type1 and Type50, each made up 50% of the total clients.

Table 5.4. Web Application Setup - Controlling 2 types of clients.

Client Type	Percentage	Initial Test Case
Type 1	50	5
Type 10	0	0
Type 25	0	0
Type 50	50	5
Type 100	0	0
Type 500	0	0
Type 1000	0	0

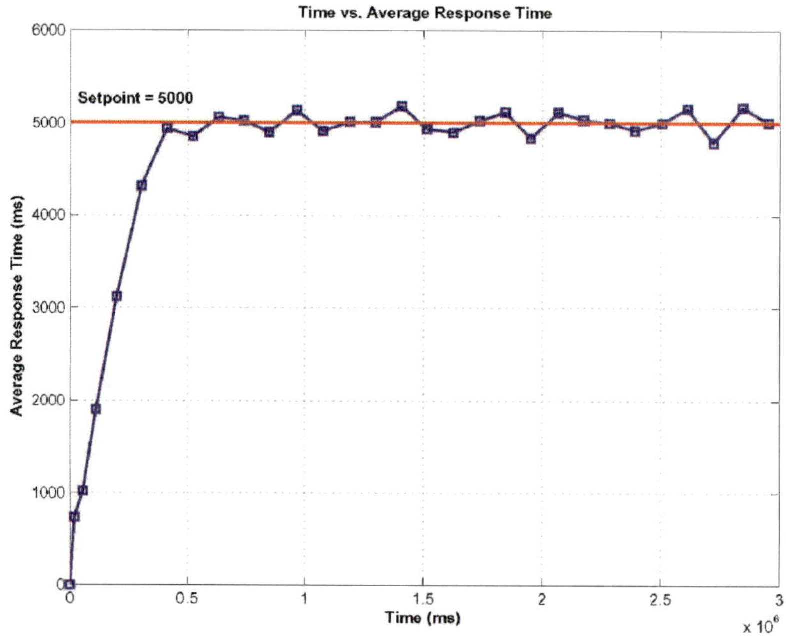

Figure 5.20. Results of controlling two types of clients (Type1 and Type50) to achieve a 5000 ms response time.

The system in Figure 5.20 started with 5 clients of Type1 and 5 clients of Type50. As time advanced and the PID Controller produced positive gains, the number of clients

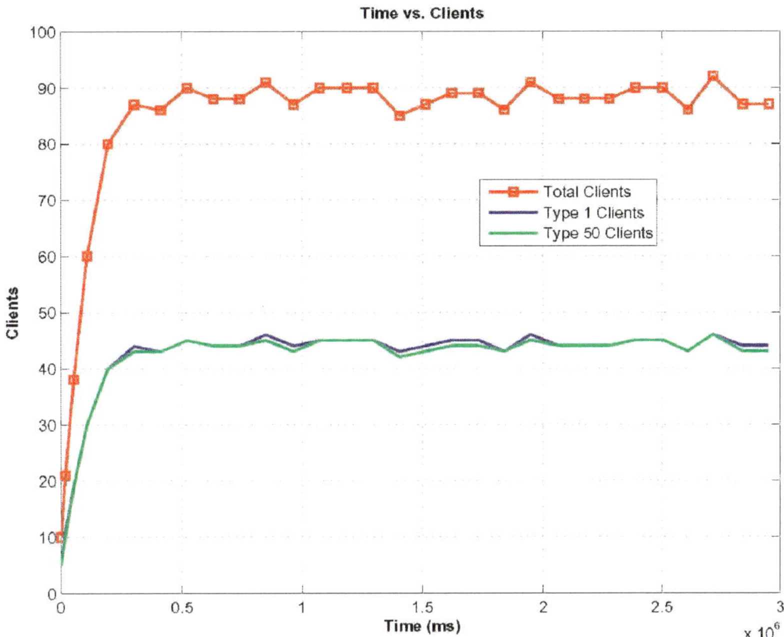

Figure 5.21. Total number of clients and number of Type1 and Type50 clients used in Figure 5.20.

increased until the Average Response Time reached the setpoint of 5000ms. After reaching the setpoint, the average response time oscillated around the setpoint.

Figure 5.21 represents the totalClients, Type1 clients, and Type50 clients in the system. As time advanced, the total number of clients increased due to the gains produced by the PID. As the total number of clients increased, Type1 and Type50 clients increased as well since each of them made 50% of the total number of clients. It required a total of 90 clients (45 of each type) to reach the desired setpoint of 5000ms.

5.4.4 Controlling Three Types of Clients

Table 5.5 represents the system setup for the third run of the experiment where 3 types of clients were controlled. The system started with a total of 5 clients of Type1. Type1 made up 34% of the total clients, Type50 made up 33% of the total clients, and Type500 made up 33% of the total clients.

Table 5.5. Web Application Setup - Controlling 3 types of clients.

Client Type	Percentage	Initial Test Case
Type 1	34	5
Type 10	0	0
Type 25	0	0
Type 50	33	0
Type 100	0	0
Type 500	33	0
Type 1000	0	0

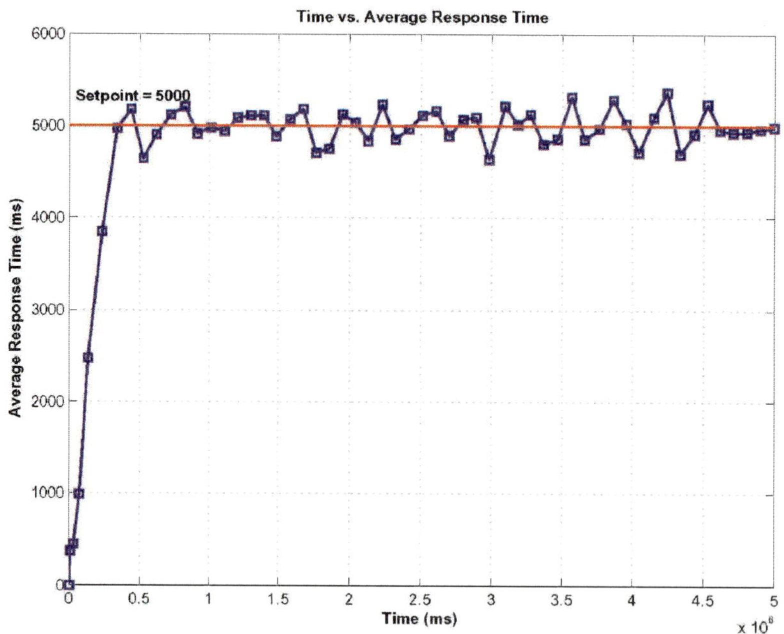

Figure 5.22. Results of controlling three types of clients (Type1, Type50 and Type500) to achieve a 5000 ms response time.

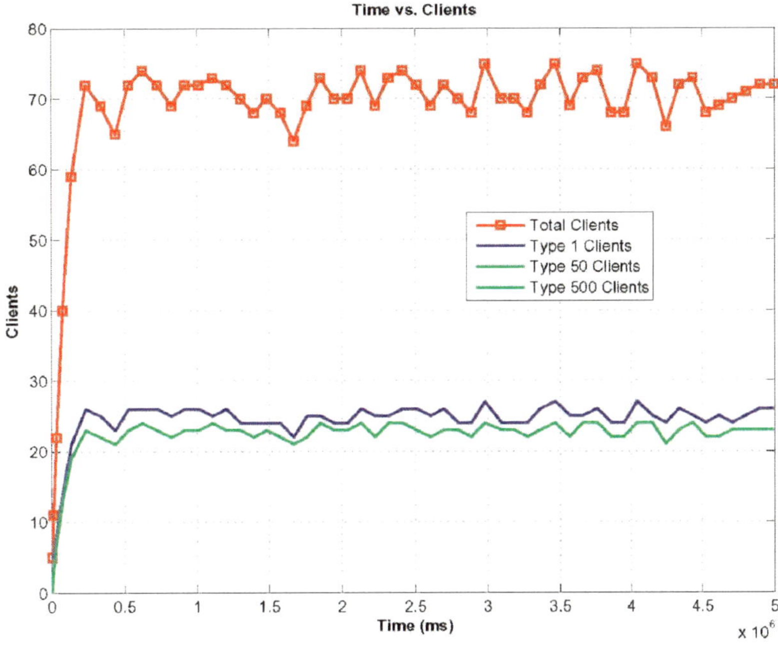

Figure 5.23. Total number of clients and number of Type1, Type50, and Type500 clients used in Figure 5.22.

The system in Figure 5.22 started with 5 clients of Type1. As time advanced and the PID Controller produced positive gains, the number of clients increased until the Average Response Time reached the setpoint of 5000ms. After reaching the setpoint, the average response time oscillated around the setpoint.

Figure 5.23 represents the totalClients, Type1, Type50, and Type500 clients in the system. As time advanced, the total number of clients increased due to the gains produced by the PID. As the total number of clients increased, Type1, Type50, and Type500 clients increased as well since each of them made about 33% of the total number of clients. It required a total of 69 clients (25 of Type1, 22 of Type50, and 22 of Type500) to reach the desired setpoint of 5000ms.

5.4.5 Controlling Four Types of Clients

Table 5.6 represents the system setup for the fourth run of the experiment where 4 types of clients were controlled. The system started with a total of 5 clients of Type1. Each of Type1, Type50, Type100, and Type500 clients made up 25% of the total clients.

Table 5.6. Web Application Setup - Controlling 4 types of clients.

Client Type	Percentage	Initial Test Case
Type 1	25	5
Type 10	0	0
Type 25	0	0
Type 50	25	0
Type 100	25	0
Type 500	25	0
Type 1000	0	0

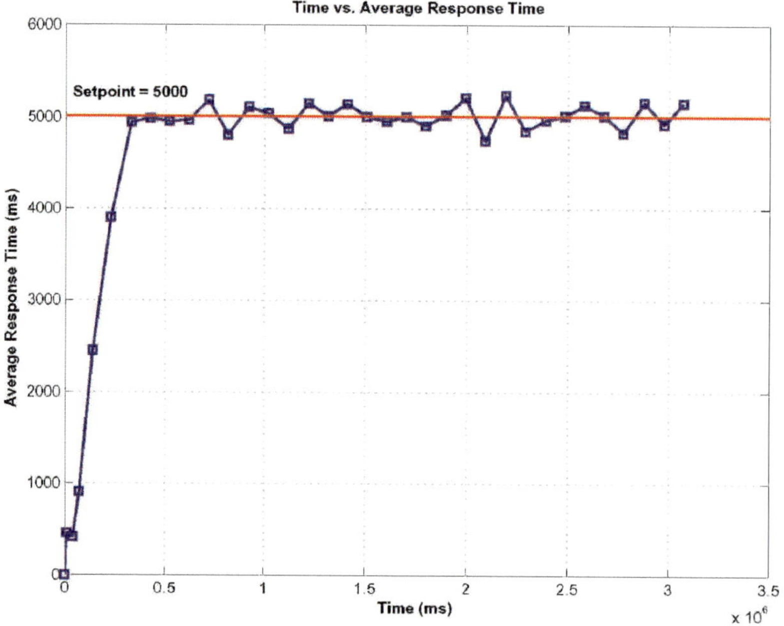

Figure 5.24. Results of controlling four types of clients (Type1, Type50, Type100 and Type500) to achieve a 5000 ms response time.

The system in Figure 5.24 started with 5 clients of Type1. As time advanced and the PID Controller produced positive gains, the number of clients increased until the Average

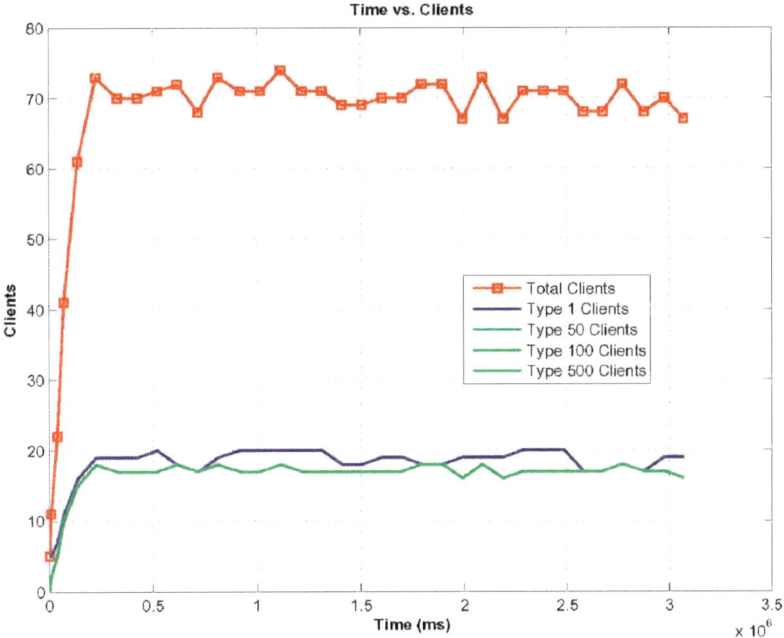

Figure 5.25. Total number of clients and number of Type1, Type50, Type100, and Type500 clients used in Figure 5.24.

Response Time reached the setpoint of 5000ms. After reaching the setpoint, the average response time oscillated around the setpoint.

Figure 5.25 represents the totalClients, Type1, Type50, Type100, and Type500 clients in the system. As time advanced, the total number of clients increased due to the gains produced by the PID controller. As the total number of clients increased, Type1, Type50, Type100, and Type500 clients increased as well since each of them made 25% of the total number of clients. It required a total of 71 clients (20 clients of Type1, 17 clients of Type50, 17 clients of Type100, and 17 clients of Type500) to reach the desired setpoint of 5000ms.

5.4.6 Controlling Seven Types of Clients

Table 5.7 represents the system setup for the fifth run of the experiment where 7 types of clients were controlled. The system started with a total of 5 clients of Type1. Each of Type1, Type10, Type25, Type50, Type100, and Type500 clients made up 15% of the total clients. Type1000 made up 10% of the total clients.

Table 5.7. Web Application Setup - Controlling 7 types of clients.

Client Type	Percentage	Initial Test Case
Type 1	15	5
Type 10	15	0
Type 25	15	0
Type 50	15	0
Type 100	15	0
Type 500	15	0
Type 1000	10	0

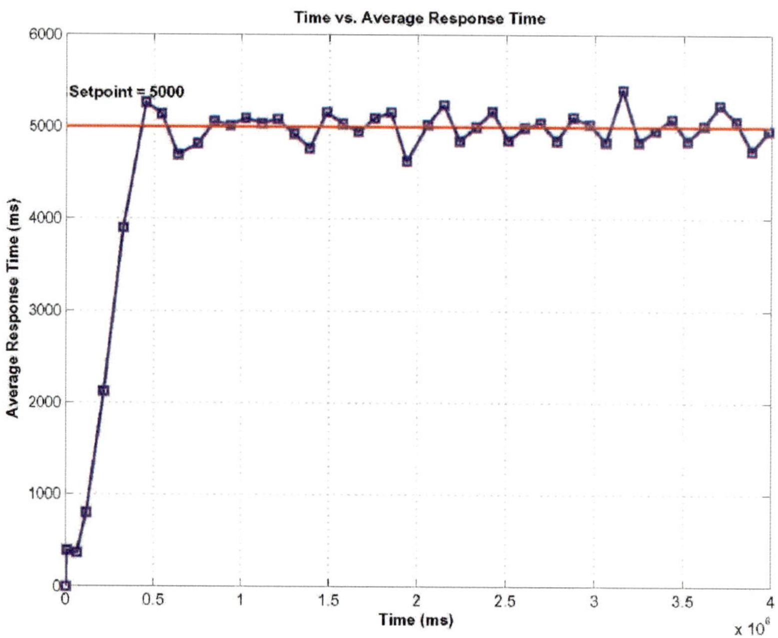

Figure 5.26. Results of controlling all seven types of clients to achieve a 5000 ms response time.

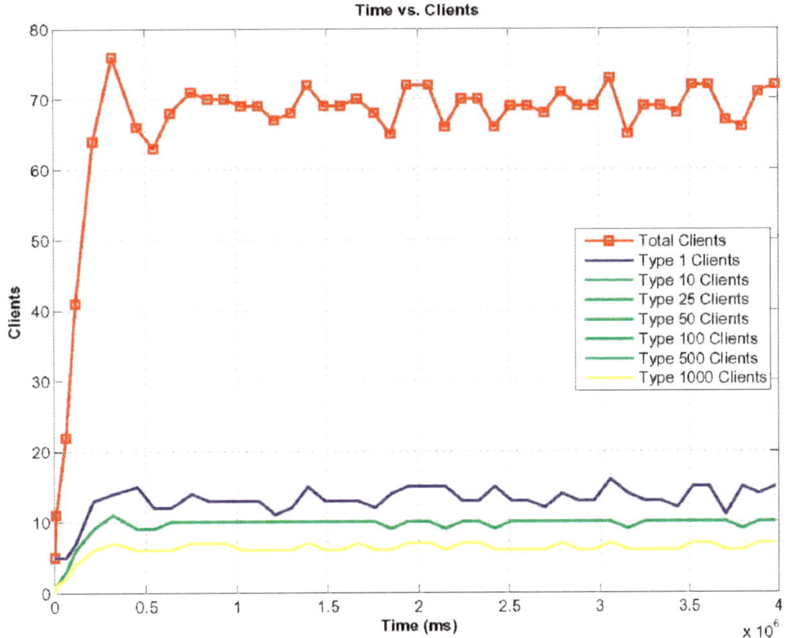

Figure 5.27. Total number of clients and number of all seven types of clients used in Figure 5.26.

The system in Figure 5.26 started with 5 clients of Type1. As time advanced and the PID Controller produced positive gains, the number of clients increased until the Average Response Time reached the setpoint of 5000ms. After reaching the setpoint, the average response time oscillated around the setpoint.

Figure 5.27 represents the totalClients and all 7 types of clients in the system. As time advanced, the total number of clients increased due to the gains produced by the PID. As the total number of clients increased, all types of clients increased as well. It required a total of 70 clients (13 of Type1, 10 of Type10, 10 of Type25, 10 of Type50, 10 of Type100, 10 of Type500, and 7 of Type1000) to reach the desired setpoint of 5000ms.

5.5 Conclusion

The four instrumented/implemented experiments are a clear indication of the high applicability and accuracy of the proposed approach. The first experiment showed that a PID

Controller can be indeed used for both Stress Testing and Load Testing of one resource with one input variable. It controlled the amount of memory an application used. Additionally, analyzing changes in input and memory allocation allowed for detection of memory leaks that were injected into the source code. The second experiment showed that when more than one input significantly affects the resource, all the inputs must be controlled to achieve the desired level of stress/load. The resource under control was response time. In the third experiment, the application under test is the Linux utility Zip and the resource under control is response time. It showed the accuracy of the Input Identification technique where the resource sensitive inputs were properly identified by the system-identification-based approach proposed here. The fourth experiment is the most realistic example so far where the application under test is a Web Application hosted by a JBoss Server and the controlled resource is the response time.

CHAPTER 6
RELATED WORK

Both black box and white box approaches have been proposed for Load and Stress Testing [48, 31, 15, 16, 22, 49]. In white box Load/Stress Testing (structural based), the program code is analyzed to produce the test cases. "Toward A Structural Load Testing Tool" by Yang and Pollock [48] and "Detecting Buffer Overflow" by Grosso [31] are examples of white box testing and they are described in Sections 6.1 and 6.2 respectively. In black box Load/Stress Testing (functional based), test cases are generated based on requirements specifications within an operational profile. Different black box testing techniques have been proposed for Load and Stress testing like "Deterministic Markov State Testing" in Section 6.3, "Stress Testing Real-Time Systems with Genetic Algorithms" in Section 6.4, and "Stress Testing of Multimedia Systems" in Section 6.5. Among a multitude of existing approaches [48, 31, 15, 16, 22, 49, 29, 23] the ones mentioned above are more closely related to the approach proposed here and are therefore described in more detail in the next sections. Also, Section 6.6 describes current available stress and loading testing tools. Finally, Section 6.7 presents a comparative description of the selected approach and tools and the PID based control approach.

6.1 Toward A Structural Load Testing Tool

One example of a white box approach is provided by Yang and Pollock [48]. Their technique identifies the load sensitive parts in sequential programs based on a static analysis of the code. A load sensitive part is "...one with the property that its correctness depends on the amount of input data or the length of time that the program will execute continuously" [48]. In other words, the program executes successfully when it is executed for a short time or under a small load and fails when executed under a heavy load or over a long period of time. In this approach, Yang and Pollock tries to identify load/stress sensitive faults which may occur in memory leaks, incorrect dynamic memory allocation, and incorrect use of user-defined memory allocation and deallocation routines. An example of a load sensitive fault in an object-oriented program, given by Yang and Pollock, is presented in Figure 6.1 where objects are continuously created with no object being destroyed which will eventually cause the program to fault with the program's space is filled with objects.

Yang and Pollock's approach computes the degree of load sensitivity of the program and regions of the program by computing the Load Sensitivity Index (LSI). Next, identify which regions may be load sensitive and require load testing based on the LSI. Finally, generate test data for load testing these identified regions of code. Yang and Pollock proposed an algorithm to compute the LSI for detecting load sensitivity due to memory allocation and deallocation. The algorithm is a recursive algorithm that traverses the control flow graph looking for memory allocation (malloc) and deallocation (free) instructions. $LSI(i)$ is defined as the LSI value at the program point just prior to instruction i, and $LSI(i')$ is defined as the LSI value at the program point just after instruction i. For a memory allocation, $LSI(i') = LSI(i) + s$ where instruction i is malloc(s bytes). For memory deallocation, $LSI(i') = LSI(i) - s$ where instruction i is free(s bytes). Otherwise, $LSI(i') = LSI(i)$. After the LSI has been computed for a program, the program is said to be:

- must-be load sensitive if the worst case $LSI > 0$ and the best case $LSI > 0$.
- may-be load sensitive if the worst case $LSI > 0$ and the best case $LSI \leq 0$.
- load insensitive or load irrelevant if the worst case $LSI \leq 0$ and the best case $LSI \leq 0$

6.2 Detecting Buffer Overflow

This is another white box approach for Stress Testing introduced by Grosso [31, 32]. The main goal of this approach is to detect buffer overflow and it does so in three steps. The main steps in Grosso's approach are outlined in Figure 6.2. First, it analyzes the software under test using static analysis techniques to determine the set of source code lines that could

```
class String
{
  char *str;

public:
  String() { str = new char; *str = 0; }
  String(char *);
  void print();
  ~String() { delete str; }
};


String::String(char *s)
{
  str = new char[strlen(s) + 1];
  strcpy(str, s);
}

void
String::print()
{
  str[strlen(str)] = '\0';
}

main()
{
  while(1)
  {
     String *sp = new String("hello world");

     sp->print();
  }
}
```

Figure 6.1. Example of a load sensitive fault in an object-oriented program

potentially cause a buffer overflow. Second, it uses program slicing [33] to detect the cause-ceffect relationship between program inputs and the lines identified in the previous step. Third, it uses Genetic Algorithms to generate test cases to identify buffer overflow threats. In summary, this approach combines static analysis, program slicing, and evolutionary testing to detect buffer overflow. Grosso's approach deals only with buffer overflow and not resource saturation.

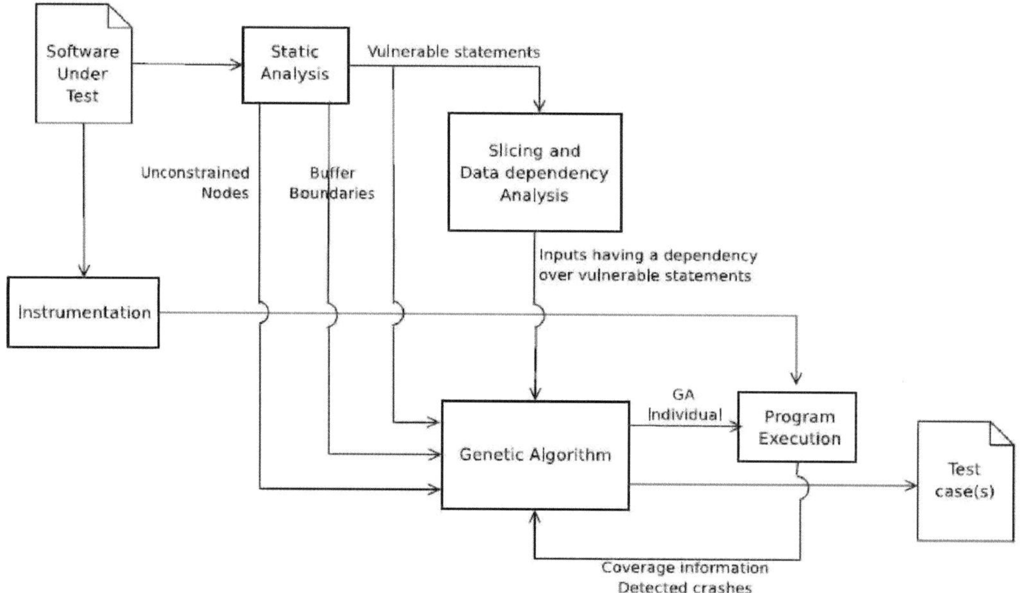

Figure 6.2. Grosso's test case generation process

6.3 Deterministic Markov State Testing

Avritzer and Larson [15, 16] proposed a black box testing technique, called Deterministic Markov State Testing, to generate test cases for testing telecommunication software applications according to the operational profiles. It is deterministic because the sequence of test case execution is set at planning time before the start of the system test, and state testing because each test case certifies a unique software state. The operational profile is used to build a Markov chain that represents the software's behavior. Only the most likely test cases, as computed from the most probable Markov chain states, are generated at planning time. Since each test case certifies a unique software state, when a failure is detected, system testers can easily trace the sequence of software states that led to the failure. This approach is limited to applications that can be modeled by using the Markov Chain because the input data is assumed to arrive according to a Poisson distribution and is serviced in an exponential distribution.

6.4 Stress Testing Real-Time Systems with Genetic Algorithms

Briand [22] proposed a technique that uses Genetic Algorithms to derive test cases to maximize the chances of deadline misses within a system. The approach, based on the system task architecture, test real-time systems that have to react to external events and internally generated system events within time constraints. The goal of the approach is to maximize the chances of critical deadline misses. Briand showed that task deadlines may be missed even though the associated tasks have been identified as schedulable through appropriate schedulability analysis where the schedulability analysis simulated the worst-case scenario of task executions.

The approach analyzes the system task architecture to determine whether deadlines can be missed. Then, it generates test cases such that the completion times of a specific tasks executions are as close as possible to their deadlines. The generated test cases concentrate on seeding times for aperiodic tasks and their possible input data since both impact task executions. Briand implemented the approach in a tool. The approach proposed here has different goals. The main concern is in resource saturation, whereas Briand's method focuses on missing deadlines. Clearly, This approach is more general than Briand's work.

6.5 Stress Testing of Multimedia Systems

To address Stress Testing in multimedia systems, Zhang [49] developed a technique and a tool to automatically generate test cases. Zhang considered a multimedia system to consist of a group of servers and clients connected by a suitable network such as video conferencing, home shopping, and remote education. Stringent timing constraints and synchronization constraints are present during the transmission of data between the servers and the clients. Zhang models the multimedia system under test using Petri-nets coupled with timing constraints and uses a specific flavor of interval temporal logic to model the temporal relationship among multimedia data within a presentation. For example, given two media objects, $VideoA$ and $VideoB$, the statement $\alpha VideoB = \beta VideoA + 5$ (where $\alpha VideoB$ is the begin time of $VideoB$ and $\beta VideoA$ is the end time of $VideoA$) is used to express the starting of $VideoB$ five time units after the end of $VideoA$. Then, Zhang extracts possible execution sequences from the Petri net. For each sequence, the event timings which maximize resource usage during a fixed finite period is computed. Basically, this algorithm divides the system statically into a set of objects where the resource usage of each object is known statically and then simulates these objects in Petri nets to compute different execution sequences.

The same goal is shared with this approach: resource saturation and testing with different loads. Zhang's technique is applicable where the specifications consist of a temporal-event Petri net and some temporal formulas for describing the relationships among media objects. This approach appropriately tackles multimedia systems where the system consists of a set of media objects and where each media object is a stand alone object with known resources requirements. Not just any system can be divided into such objects and not all systems can be or are specified using Petri nets. Also, Zhang's technique cannot be easily generalized to stress test other kinds of resources, such as network traffic as it would require important changes in the test model. In other words, Zhang's algorithm has applicability limitations. If a system can be divided into different objects with known resources requirements and these objects can be easily specified in petri nets, then it may be easier to use Zhang's technique. However, the approach in this work is applicable to any system.

6.6 Stress/Load Testing tools

When considering Performance, Load, and Stress Testing of web applications, a vast number of existing tools are available. For example, load-testing tools evaluate the performance of a specific web site on the basis of actual users' behavior; they capture real user requests for further replay where a set of test parameters can be modified. Among a multitude of existing stress/load testing tools [42, 47, 5, 2, 1, 4, 17, 37, 7, 6, 11, 10, 3, 8] five of the most popular tools, Load Runner [2], Silk Performer [7], Performance Tester [6], WAS [4], and Load Tester [3], are compared in table 6.1. Table 6.1 compares the stress/load testing tools based on the following list of functionalities:

- Virtual Users: simulating concurrent users sending requests to the server. Each virtual user is independent where it can have its own cookies, data, and other parameters.
- Script Recording: recording a testing script by walking through a site with a browser. Play back and make modifications to the script as needed.
- Distributed Load Testing: Distribute the load generation on multiple machines and control all of them from one central machine. The master client distributes test data to all clients, initiates the test in all clients simultaneously, and collects the test results and reports from all clients. For example, to run a test of 1000 users, you can use three or more machines with another master machine controlling them.
- Parameterization: allowing the tester to replace recorded values in a script with parameters. It is often used when the application needs unique data such as user name, data dependency such as passwords, data cache, or date constraints.
- Data Correlation: Data correlation handles dynamic content. Dynamic content refers to variables dynamically created during the execution of a test case and these variables may be used during a later step. This feature prevents hard coding of values.

- Test scheduling: Using a scheduler to simulate a given sequence of events. For example, the volume of transactions can be increased at defined intervals to increase the load during a test.
- Logging: logging of virtual users activities.
- Reporting: Reporting capabilities to help improve application performance.

Table 6.1. Comparison of Stress/Load Testing tools by Functionalities

Features	Stress/Load Testing Tools				
	Load Runner	Silk Performer	Performance Tester	WAS	LoadTester
Virtual Users	✓	✓	✓	✓	✓
Script Recording	✓	✓	✓	✓	✓
Distributed Load Testing	✓	✓		✓	✓
Parameterization	✓	✓	✓	✓	✓
Data Correlation	✓		✓		
Test Scheduling	✓	✓	✓	✓	✓
Logging	✓	✓	✓	✓	✓
Reporting	✓	✓	✓	✓	✓

All of the Stress/Load testing tools try to achieve load generation and performance measurement by simulating client requests to the application under test. They all require user configuration for specifying the number of clients to simulate and the test cases to run. After each run, the tester manually revisits the test cases and adds or modifies them to achieve the test load required. The approach discussed here automates the process of test loading the application. It also automatically identifies the resource sensitive inputs. To achieve this, the testers are only required to provide an initial test case. On the other hand, the approach discussed here can complement the existing tools by automating the test case creation of some existing tools. Basically, the testing tool would be the "Wrapper" in Figure 3.3.

6.7 Conclusion

White box testing techniques, Yang and Pollock's approach and Grosso's approach, are time consuming and require the testers to have certain programming skills. The approach proposed here presents a higher applicability because it does not depend on source code availability and it can be used to potentially control any type of resources. White box techniques are, in general, restricted to certain types of resources. Also, Grosso's approach deals only with buffer overflow and not resource saturation.

Avritzer and Larson's approach is similar to the approach proposed here as both concentrate on available resources. However, Avritzer and Larson's approach is limited to systems that can be modeled by Markov Chain, and it generates the test case before starting the system test. The approach proposed here is more general, and it generates the test cases automatically. Briand's approach has different goals from the approach proposed here where the main concern is missing deadlines, whereas the approach proposed in this work focuses on resource saturation. Clearly, This approach is more general than Briand's work. Zhang's algorithm has applicability limitations as it only applies to systems can be specified using Petri nets. A vast number of Stress/Load testing tools are available but none of them automate the testing process like the work proposed here.

In terms of accuracy, if the system is controllable and stable or asymptotically stable [41], the use of control theory guarantees that the stress/load level will be achieved. If the system is uncontrollable and unstable, no technique can be accurate. However, the approach proposed in this work depends on proper identification of the inputs affecting the resource. As described in Chapter 2, use of system identification techniques [40] have shown to be reasonably accurate to identify such inputs.

Table 6.2 presents a summary/comparison between the approach proposed here and other related approaches. The comparison is based on four key features. First, whether the approach is widely applicable to different types of applications. Second, if the approach automatically generates test cases and reaches the desired load level. Third, whether its applicable to a wide range of resources like memory usage, and response time. Finally, does the approach adjusts its test cases to maintain a desired load level regardless of the testing environment and any dynamic changes in the operating environment.

Table 6.2. Comparing the proposed work with other approaches

Approach	Widely Applicable	Automatic	Flexible	Adjust to Evn. Changes
Pollock	✓	✓		
Grosso	✓			
Avritzer		✓	✓	
Briand		✓		
Zhang				
Stress Testing Tools			✓	
Proposed Approach	✓	✓	✓	✓

CHAPTER 7

CONCLUSION AND FUTURE WORK

7.1 Conclusion

Computer Science and Software Engineering fields have been too restricted to discrete solutions and traditional continuous approaches have not been used widely. While the time interval for continuous systems may be nano seconds, it can be seconds or even tens of seconds for software applications. PID controllers are most often used in continuous processes where the output is a continuous flow like a chemical process or a refining process for gasoline. However, PID control theory can also be used in a discrete manner and the work proposed here exploits the potential of using a simple control technique like a PID Controller to achieve what other discrete approaches have failed to deliver.

In this work, a Stress and Load testing technique is proposed based on rigorous theories, feedback control theory and system-identification theory. It is an automated technique that exploits the advantages of controller tuning to efficiently and accurately control a system using PID controller. The approach, though of special interest for embedded systems and web servers, is not restricted to any one type of system or application. The approach can be used to automatically control any type of resource (as long as it can be measured) and can also accurately identify resource-sensitive inputs in an automated manner. Existing approaches present limitations in terms of applicability and resources to be controlled. A more general approach is offered in this work.

The work presented here has four major contributions. First, an Input Identification algorithm is proposed to identify resource-sensitive inputs in an automated manner. The algorithm is based on System Identification techniques. Second, the proposed work exploits PID control theory to control and automate stress and load testing. Third, a PID tuning technique appropriated for software systems has been developed. The fourth major contribution is that the approach, unlike existing approaches, does not present limitations in terms of applicability where it can be applied to a wide range of software applications as long as they are properly defined in terms of inputs and outputs and as long as the resource under control is measurable (like memory or response time).

Extensive work was done to validate the applicability of the approach. Four experiments were conducted to show the potential and accuracy of the proposed work. The first experiment, examined control of the amount of memory an application used where the resource under control is memory. The second experiment, showed that when more than one input significantly affects the resource, all the inputs must be controlled to achieve the desired level of stress/load. The resource under control was response time. In the third experiment, the application under test is the Linux utility Zip and the resource under control is the response time. It showed the accuracy of the Input Identification technique where the resource sensitive inputs were properly identified by the system-identification-based approach proposed here. The fourth experiment showed the applicability of the approach to the most widely used software applications, web applications. Also, in the fourth experiment, the resource under control is the response time. In summary, the results of experiments indicated that the approach proposed in this work offers both high applicability and accuracy.

7.2 Future Work

The work proposed here consisted of three major problems. Automatic identification of resource sensitive inputs, automatic Stress/Load Testing for a resource using PID control theory, and automatic PID tuning technique appropriated for software systems have been handled in this work. This proposed work dealt with controlling one resource at a time such as memory or response time. In other words, so far, the work used a PID controller which is a SISO (Single Input Single Output) controller to change each resource-sensitive input parameter based on the gain produced by the controller. In the scenarios where there were multiple inputs affecting the resource, a set of independent SISO controllers were used. In some scenarios, stress/load testing may require controlling more than one resource at the same time like memory and disk space and some inputs may affect both resources in positive or negative manners. For example, a system may have $\{I_1, I_2, \ldots, I_{15}\}$ as the set of inputs while inputs $\{I_4, I_7, I_8, I_{11}\}$ significantly affect resource R_1 and inputs $\{I_1, I_4, I_8, I_{12}\}$ significantly affect resource R_2. Input I_4 affects R_1 in a positive manner but it affects R_2 in a negative way. On the other hand, input $\{I_8\}$ affects both Resources R_1 and R_2 in a positive way. So, if there is a SISO controller increasing I_8 to achieve a certain level of usage for R_1, it will increase the usage of R_2 as well. On the other hand, if there is another SISO controller controlling I_4 to achieve a certain level of usage for R_2, the two controllers affect each other and are said to be interacting.

The problem of interaction can be minimized with the addition of a decoupling element in the control system, although this further increases the complexity. For example an n by n process requires a total of n PID controllers and n decouplers. Also, tuning a number of PID controllers can be difficult and time consuming. Another solution would be to use a centralized multivariable controller which is used to control MIMO (Multiple Inputs Multiple Outputs) processes. In Chapter 2, a state-space model was generated, using System Identification, which represents the system under test (MIMO system). Currently, the generated state-space model is used to tune the PID controller. However, it can also be used to build a multivariable controller to control the MIMO system and achieve the desired level of stress/load testing. The plan for future work is to examine different types of control as well as multi-variable control techniques based on the generated state-space model to control multiple resources at the same time.

Another issue to tackle is the stability analysis of the system. The stability analysis is needed to identify if the system is unstable and therefore cannot be controlled. In general, a system is unstable if the system response approaches infinity as time approaches infinity; if G(t) is the system, then it is unstable if $\lim_{t\to\infty} \|G(t)\| = \infty$. Lyapunov stability [13] deals with the stability of dynamical systems. A system is Lyapunov stable, if all solutions of the dynamical system that start close enough to equilibrium point x_e stay near x_e forever. Moreover, if the system is Lyapunov stable and all solutions that start close enough to x_e eventually converge to x_e, then it is asymptotically stable. Furthermore, if the system is asymptotically stable and guarantees a minimal rate of decay, then it is exponentially stable. Using Lyapunov stability, a system can be determined whether it can be controlled or not.

REFERENCES

[1] The grinder, a java load testing framework. http://grinder.sourceforge.net.

[2] Loadrunner by mercury interactive. http://www.mercuryinteractive.com.

[3] Loadtester by appperfect. http://www.appperfect.com/products/load-test.html.

[4] Microsoft web application stress tool. http://www.microsoft.com/.

[5] Opensta (open system testing architecture). http://www.opensta.org.

[6] Rational performance tester by ibm. http://www-01.ibm.com/software/awdtools/tester/performance/.

[7] Silkperformer by borland. http://www.borland.com/us/products/silk/silkperformer/.

[8] Sitetester by pilot software. http://www.pilotltd.com/.

[9] Star telegram. http://www.star-telegram.com/2010/04/07/2097615/texas-23-million-in-appliance.html.

[10] Wapt by softlogica. http://www.loadtestingtool.com/.

[11] Webserver stress tool by paessler. http://www.paessler.com/webstress.

[12] Zip reference manual pages. http://www.linuxmanpages.com/man1/zip.1.php, August 1999.

[13] Ling Hou Anthony N. Michel and Derong Liu. *Stability of Dynamical Systems: Continuous, Discontinuous, and Discrete Systems.* Systems and Control: Foundations and Applications. Birkhuser Boston, 2008.

[14] Jeffrey Arbogast, Douglas J. Cooper, and Robert C. Rice. Model-based tuning methods for pid controllers. Control Station, Inc.

[15] Alberto Avritzer and Brian Larson. Load testing software using deterministic state testing. In *Proceeding of the International Symposium on Software Testing and Analysis*, pages 82–88. ACM, June 1993.

[16] Alberto Avritzer and Elaine J. Weyuker. Generating test suites for software load testing. In *Proceedings of the International Symposium on Software Testing and Analysis*, pages 44–57, Washington, Seattle, August 1994. ACM.

[17] Gaurav Banga and Peter Druschel. Measuring the capacity of a web server. In *USENIX Symposium on Internet Technologies and Systems*, 1997.

[18] M. Bayan and J. W. Cangussu. Automatic feedback, control-based, stress and load testing. In *Proceedings of the 23rd Annual ACM Symposium on Applied Computing*, Fortaleza, Ceara, Brazil, March 2008. ACM.

[19] Boris Beizer. *Software System Testing and Quality Assurance.* Van Nostrand Reinhold, 1984.

[20] Boris Beizer. *Software Testing Techniques.* International Thomson Computer Press, 2nd Edition, 1990.

[21] Robert V. Binder. *Testing Object-Oriented Systems - Models, Patterns, and Tools.* Addison-Wesley, 1999.

[22] Lionel C. Briand, Yvan Labiche, and Marwa Shousha. Stress testing real-time systems with genetic algorithms. In *Proceedings of the Genetic And Evolutionary Computation Conference*, pages 1021–1028, Washington, DC, June 2005. ACM.

[23] Giuliano Casale, Amir Kalbasi, Diwakar Krishnamurthy, and Jerry Rolia. Automatic stress testing of multi-tier systems by dynamic bottleneck switch generation. In *Mid-*

dleware '09: Proceedings of the 10th ACM/IFIP/USENIX International Conference on Middleware, pages 1–20, New York, NY, USA, 2009. Springer-Verlag New York, Inc.

[24] G. H. Cohen and G. A. Coon. Theoretical consideration of related control. In *Transactions of the ASME*, pages 827–834, 1953.

[25] M. Dash and H. Liu. Feature selection for classification. *Intelligent Data Analysis*, 1:131–156, 1997.

[26] B. De Schutter. Minimal state-space realization in linear system theory: an overview. *J. Comput. Appl. Math.*, 121(1-2):331–354, 2000.

[27] Federal standard 1037c. General Services Administration Information Technology Service, August 1996. Prepared by: National Communications System Technology and Standards Division.

[28] William A. Florac and Anita D. Carleton. *Measuring the Software Process: Statistical Process Control for Software Process Improvement.* SEI Series in Software Engineering. Addison-Wesley, Reading , MA, 1999.

[29] Vahid Garousi, Lionel C. Briand, and Yvan Labiche. Traffic-aware stress testing of distributed systems based on uml models. In *ICSE '06: Proceedings of the 28th international conference on Software engineering*, pages 391–400, New York, NY, USA, 2006. ACM.

[30] Graham C. Goodwin, Stefan F. Graebe, and Mario E. Salgado. *Control system design*. Prentice Hall, Upper Saddle River, New Jersey, 2001.

[31] Concettina Del Grosso, Giuliano Antoniol, Massimiliano Di Penta, Philippe Galinier, and Ettore Merlo. Improving network applications security: a new heuristic to generate stress testing data. In *Proceedings of the Genetic and evolutionary computation Conference*, pages 1037–1043, Washington, DC, 2005. ACM.

[32] Concettina Del Grosso, Supervised Giuliano Antoniol, and Massimiliano Di Penta. An evolutionary testing approach to detect buffer overflow. In *In Student Paper Proceedings of the International Symposium of Software Reliability Engineering (ISSRE*, pages 77–78, 2004.

[33] Mark Harman and Keith Brian Gallagher. Program slicing. In *Software Focus*, pages 70–79. IEEE Computer Society Press, 1981.

[34] Ieee standard glossary of software engineering terminology. IEEE Standards Collection - Software Engineering, 2002.

[35] A. Jain and D. Zongker. Feature selection: evaluation, application, and small sample performance. In *Pattern Analysis and Machine Intelligence, IEEE Transactions on*, pages 153–158. IEEE, Feb. 1997.

[36] Jer-Nan Juang. *Applied System Identification*. Prentice Hall, Englewood Cliffs, New Jersey, first edition, 1993.

[37] Krishna Kant, Vijay Tewari, and Ravi Iyer. Geist: A generator for e-commerce internet server traffic. In *IEEE International Symposium on Performance Analysis of Systems and Software (ISPASS)*, pages 49–56. IEEE, 2001.

[38] K. Kira and L.A. Rendell. The feature selection problem: Traditional methods and a new algorithm. In *Ninth National Conference on Artificial Intelligence*, page 129134, 1992.

[39] Daphne Koller and Mehran Sahami. Toward optimal feature selection. pages 284–292. Morgan Kaufmann, 1996.

[40] Lennart Ljung. *System Identification: Theory for the user*. Prentice-Hall, Englewood Cliffs, New Jersey, 1987.

[41] David G. Luenberger. *Introduction to Dynamic Systems: Theory, models and applications*. John Wiley & Sons, New York, 1979.

[42] David Mosberger and Tai Jin. httperf – a tool for measuring web server performance. *SIGMETRICS Perform. Eval. Rev.*, 26(3):31–37, 1998.

[43] P.M. Narendra and K. Fukunaga. A branch and bound algorithm for feature subset selection. *Computers, IEEE Transactions on*, C-26(9):917–922, Sept. 1977.

[44] John A. Shaw. Pid algorithms and tuning methods. URL: http://www.jashaw.com/pid/tutorial.

[45] T. Wescott. Pid without a phd. Technical report, Embedded System Programming, 2000.

[46] Donald J. Wheeler and David S. Chambers. *Understanding Statistical Process Control.* SPC Press, Knoxville, Tennessee, 1992.

[47] Katinka Wolter and Kristian Kasprowicz. Webapploader: A simulation tool set for evaluating web application performance. In *Computer Performance Evaluation / TOOLS*, pages 47–62, Sept. 2003.

[48] Cheer-Sun D. Yang and Lori L. Pollock. Towards a structural load testing tool. In *Proceedings of the International Symposium on Software Testing and Analysis*, pages 201–228, San Diego, CA, 1996.

[49] Jian Zhang and S. C. Cheung. Automated test case generation for the stress testing of multimedia systems. *Softw., Pract. Exper.*, 32(15):1411–1435, 2002.

[50] J. G. Ziegler and N. B. Nichols. Optimum settings for automatic controllers. In *Trans. Amer. Soc. Mech. Eng.*, pages 759–768, 1942.

VITA

Mohamad Bayan was born in Baalbeck, Lebanon on October 27, 1979. After finishing high school from the Secondary Evangelical School, Zahle, Lebanon, Mr. Bayan enrolled at The University of Tennessee at Chattanooga, Chattanooga, TN for one year before transfering to The University of Texas at Dallas, Richardson, TX where he received an Academic Excellence Scholarship. Mr. Bayan completed the Bachelor of Science in Computer Science from The University of Texas at Dallas in August 2003. Mr. Bayan received the degree of Master of Science in Computer Science with Major in Software Engineering in December 2005 and the Doctoral Degree in Computer Science in August 2010 from The University of Texas at Dallas. His general research interests include exploiting Control theory for the automation of tasks in Software Engineering, his current research activity is in the automation of Stress and Load Testing through automatic creation of test cases based on resource performance level.

Mr. Bayan worked for Nortel Networks in Richardson, TX as a CO-OP Student from May 2004 to May 2005. Mr. Bayan also worked as a Research Assistant at the University of Texas at Dallas from September 2005 to December 2005. Since January 2006, Mr. Bayan has been employed as a Software Engineer at Crossvale Inc. in Plano, TX.

CPSIA information can be obtained
at www.ICGtesting.com
Printed in the USA
LVIW011106111112
306812LV00005B